MULTIDIMENSIONAL FILTER BANKS AND WAVELETS
Research Developments and Applications

edited by

Sankar Basu
IBM Thomas J. Watson Research Center
Bernard Levy
University of California, Davis

A Special Issue of
**MULTIDIMENSIONAL SYSTEMS
AND SIGNAL PROCESSING**
An International Journal
Volume 8, Nos. 1/2 (1997)

KLUWER ACADEMIC PUBLISHERS
Boston / Dordrecht / London

MULTIDIMENSIONAL SYSTEMS AND SIGNAL PROCESSING

An International Journal

Volume 8, No. 1/2, January 1997

Special Issue: Multidimensional Filter Banks and Wavelets: Research Developments and Applications
Guest Editors: Sankar Basu and Bernard Levy

Distributors for North America:
Kluwer Academic Publishers
101 Philip Drive
Assinippi Park
Norwell, Massachusetts 02061 USA

Distributors for all other countries:
Kluwer Academic Publishers Group
Distribution Centre
Post Office Box 322
3300 AH Dordrecht, THE NETHERLANDS

ISBN 978-1-4419-5171-7

Library of Congress Cataloging-in-Publication Data

A C.I.P. Catalogue record for this book is available
from the Library of Congress.

Multidimensional Systems and Signal Processing, 8, 5–6 (1997)

Editorial

The Special Issue on **Wavelets and Multiresolution Analysis** is a complete set of two volumes on recent advances in subband coding, wavelets, and multiresolution analysis. The first volume has already been published under the title, "Multidimensional filter banks and wavelets—Part I: Basic theory and cosine modulated filter, banks," by combining issues 3 and 4 of Volume 7 of this journal. This is the second volume which appears as "Multidimensional filter banks and wavelets—Part II: Recent developments and some applications," generated by combining the first two issues of volume 8. Overall, the two volumes comprising the Special Issues contain six papers devoted to wavelets and subband coding theory, four papers on applications, and one regular paper plus a communication brief concerned primarily with computational tasks. A nice balance of contributions from universities and industries is also noticeable. The simultaneous attention directed towards tutorial as well as original research aspects should be attractive to non-specialists as well as specialists. Realizing the expected impact and value to researchers and scientists, these joint issues will be available as hardcover volumes for greater convenience in accessing and acquisition. The importance and timeliness of the subject-matter is further substantiated by the recent appearance of a Proceedings of IEEE Special Issue on wavelets, guest edited by Ingrid Daubechies and Jelena Kovacevic (who coauthored two papers in the first volume referred to above). The two Special Issues are somewhat different and complement each other in the sense that while the one published by Proceedings of IEEE has a slightly broader appeal, the other published by this journal is more focussed, especially on proven and potential multidimensional generalizations.

Since the inauguration of the journal in March 1990, there have been four Special Issues, including this one, devoted to topics that are central to the aims and scopes of this journal. At least one more, announced in an earlier editorial, is in an advanced planning stage, while a new one under the working title, **New Applications for 2D Systems Control Theory**, is in a preliminary stage of discussion under the initiative of Dr. Eric Rogers, an associate editor for this journal. If carried to timely completion, the readers can hope to read about exotic applications concerned with, for example, two-dimensional self-tuning control of paper-making supported by adequate theory.

There is a small change in the composition of our Editorial Board members. Dr. Y. Genin had expressed a desire to step aside because of his changing research interests. In deference to his earlier contributions, I had requested him to stay on for a while until a suitable replacement could be identified. Dr. Petre Stoica of Uppsala University has now agreed to assume that role. So, I take this opportunity to thank Yves Genin for his past contributions and welcome Petre Stoica as the newest member of our Editorial Board. More changes will be forthcoming for infusion of new ideas and initiatives and suggestions are invited.

N. K. Bose
Editor-in-Chief

The following papers appeared in issue 7:3/4 of MULTIDIMENSIONAL SYSTEMS AND SIGNAL PROCESSING. Together with the papers in this issue, this body of work represents a total of four issues developed by Guest Editors Sankar Basu and Bernard Levy. We are grateful for their contributions to MULTIDIMENSIONAL SYSTEMS AND SIGNAL PROCESSING.

Yuan-Pei Lin and P. P. Vaidyanathan - Theory and Design of Two-Dimensional Filter Banks: A Review

R. Bernardini and J. Kovacevic - Local Orthogonal Bases I: Constructions

R. Bernardini and J. Kovacevic - Local Orthogonal Bases II: Window Design

Multidimensional Systems and Signal Processing, 8, 7–10 (1997)

Preface

This is a companion volume of the first part of a two-part special issue on Wavelets and Multiresolution Analysis recently published by the journal Multidimensional Systems and Signal Processing. Although the two parts are published under separate covers, and can be read independently, the reader will better appreciate organization of the material if the two volumes are used together. While the first part contained an extensive review of the status of multidimensional perfect reconstruction (PR) filter banks and wavelets by Lin and Vaidyanathan, and a two part paper on the extension of local orthogonal bases by Bernardini and Kovacevic to multidimensions, the present volume contains a larger number of papers, emphasizing both theory and applications of multidimensional wavelets and filter banks. Thus, in the present volume, we have tried to group the papers into several loosely tied categories covering Theory, Applications and Computation.

As interest in the field continues to expand, we have witnessed announcements of a number of special issues on wavelet related topics since the preface for the first part was written. Among these, a special issue of the Proceedings of the IEEE edited by Kovacevic and Daubechies has appeared almost concurrently with the publication of the first part of this special issue, and we are aware of at least two other special issues announced by IEEE that are due to appear early next year. The former has a very broad coverage consisting of impact of wavelets in science and engineering in general, and consists of papers primarily of tutorial nature.

As announced in the preface of the first part, one of our goals in editing this special volume was to focus on multidimensional problems alone, and to demonstrate that many results on wavelets and multiresolution analysis fit in the traditional framework of multidimensional systems and signal processing. As the field matures, it appears that the remaining unresolved theoretical problems are largely multidimensional in nature. These include, among many others, the design of perfect reconstruction filter banks with circularly symmetric frequency responses (sometimes called isotropic filters in the wavelet community), wavelets for nonuniformly sampled signals in the plane and so on.

On the other hand, it has been argued that most signal processing problems are inherently multidimensional, but are considered after collapsing them to one of several dimensions. An example being acoustic signals which truely exist in three spatial and one temporal dimension, but are routinely processed as one-dimensional signals (there also exist situations where the opposite is true e.g., speech, which is primarily one-dimensional, is often studied via a large number of very high dimensional feature vectors for a variety of purposes including recognition).

Although the number of publications on wavelets has been increasing at an exponential rate, our hope is that the papers collected in these two volumes fit into a coherent theme, and the readers will be able to see a certain point of view.

Brief Summary of the Articles

This second part opens with a set of three more theory papers, which begins with a contribution by Park, Kalker and Vetterli on the use of Groebner basis in wavelet construction. Groebner basis methods have traditionally played an important role in multidimensional signal and system theory. The primary reason is that it provides a computational tool for solving the so called Bezout identity, which in the multidimensional context, can be viewed as equations arising from the Hilbert's Nullstellensatz. The paper by Park, Kalker and Vetterli entitled *Groebner Basis and Multidimensional FIR Multirate Systems* exemplifies the usefulness of the Groebner Basis technique in the multidimensional wavelet context by essentially considering the perfect reconstruction filter bank design problem as an algebraic problem. While only the surface of an entire problem area is scratched in this ground breaking paper, we can foresee a larger body of work along these lines in the near future. Many perfect reconstruction filter bank design problems formulated in this way have close connections with the celebrated Quillen-Suslin theorem and its variants. Open problems abound in this area, and include PR linear phase FIR filter bank design, design of PR multidimensional filter banks, which are stable and causal at both analysis and synthesis ends to mention a few. While available results are so far only preliminary, and more questions can be asked than has been answered in the paper included in this special issue, such as the questionable utility of Groebner basis design methods in the filter bank context due to alleged numerical instability of Groebner basis computations in finite precision arithmetic, it is believed that much research will focus on related issues on this specific topic in the near future.

Multiwavelet is a relatively new twist in wavelet theory. Work in this area has been spurred by conflicting demands on wavelet properties, and so far has been primarily motivated by the need for shorter filters, as well as to establish closer ties with certain aspects of theory of splines. While a comprehensive understanding of the theory of multiwavelets, and their usefulness in practical problems is far from complete at this point of time, the paper of Micchelli and Xu, *Reconstruction and Decomposition Algorithms for Biorthogonal Multiwavelets*, builds on previous work of the same authors and offers an approach to the construction of multiwavelets and the ensuing biorthogonal wavelet bases on bounded domains. Biorthogonal multiwavelets in a weighted Hilbert space are constructed. The wavelets are generated recursively starting on the coarsest level. These initial wavelets result from a matrix completion problem combined with suitable subsequent matrix transformations. The constructed wavelets are discontinuous but are claimed to be useful for solutions of integral equations. A concrete univariate example with the Chebyshev weight function is described to show the construction. While the matrix formulation of the decomposition and reconstruction algorithms described here also fits into the framework used by Wolfgang Dahmen and collaborators, there exists other approaches to studying multiwavelets (e.g., recent work by Strella and Strang, Geronimo and Harden). Relationship of the approach of Xu and Micchelli with the work of the others still remains to be explored.

The paper by Marshall on *Zero Phase Filter Bank and Wavelet Code Matrices: Properties, Triangular Decompositions, and a Fast Algorithm* is again algebraic in flavour, and many in depth issues here are closely related to the earlier paper by Park, Kalker and Vetterli. Here, discussion starts with one-dimension, and the author shows how Daubechies wavelets and the so called "coiflets" can be derived via a group theoretic formulation of the PR filter bank design problem. Extensions to two-dimensions are also partially achieved for the special case of McClellan transform based designs. Ladder filter structure, which is known to be closely related to the Euclidean division algorithm, again features here as an important ingredient.

The next group of four papers were definitely motivated by applications of wavelets in multidimensional signal processing. In the paper *On the Translation Invariant Subspaces and Critically Sampled Wavelet Transforms* by Benno and Moura, the authors consider the problem of lack of translation invariance of wavelet transforms, which has been a bottleneck in many practical applications of wavelet theory including in radar, sonar, pattern recognition of images etc. In the present paper, the problem is addressed by assessing the degree of translation invariance of a wavelet by the percentage of the energy preserved by the wavelet subspace at a certain scale when the input signal is arbitrarily shifted. The authors show that the variation in the said energy can be estimated by using the Zak transform of the mother wavelet. Two methods are also presented to improve the translation invariance of the wavelet basis. The approach may be contrasted with other approaches which involve finding the 'best shift' representation from a library of shifted basis functions or approaches essentially involving oversampled representations. While these latter approaches may be appropriate for image coding, the present approach is motivated by detection problems in multipath environments characteristic of seismic or sonar applications.

Image compression has been one of the major driving forces behind the recent proliferation of interest in wavelets. Attempts to compare and/or consolidate wavelet type techniques in JPEG and MPEG standards exemplifies this trend. While image compression is a vast subject by itself, and any attempt at completeness even within the limited scope of this special issue would have been futile, the papers *Low Bit-rate Design Considerations for Wavelet-Based Image Coding* by Lightstone, Majani, Mitra and *Multiresolution Vector Quantization for Video Coding* by Calvagno and Rinaldo would provide the reader some flavour of how subband coding and wavelets are currently being used in compression of still pictures and video images.

Next, in a very comprehensive paper entitled *Multiscale statistical Anomaly Detection: Analysis and Algorithms for Linearized Inverse Scattering Problems*, Miller and Willsky discuss the use of wavelet based techniques in diffraction tomography. While we have categorized it as an applications paper, much theory underlies the considerations involved, and the paper can be seen as an excellent example of how theory of wavelets can be put to novel applications such a ultrasonic, medical imaging including geophysical prospecting via its use as a tool for solving inverse scattering problems. It is, perhaps, fair to remark that use of wavelets in a estimation theoretic framework, as is the case in this paper, is less common and relatively more recent. Born inverse scattering technique from diffraction tomography is used and the all too often mentioned ability of the wavelet transforms to 'zoom in' or 'zoom out' comes handy in detection of local anomalies.

Attention shifts to computational paradigms in the following paper by Fridman and Manolakos entitled *On the Scalability of the 2-D Wavelet Transform Algorithms*. Scalable computing is an important and evolving area of research in general, and the authors show how wavelet computations nicely fit into this type of computational methodology. Close ties between wavelet techniques and the multigrid method for solving PDEs are now well known. Readers familiar with this synergy may find this paper interesting or useful due to the massive computation demanded by PDE solvers. Essentially, here, scalability results form the fact that the wavelet transform, in one or in multiple dimensions, can be viewed as repeated convolution and decimation operations.

A problem of interest in computer graphics, translates to estimating the inverse of a real valued function, of which only the numerical values over a lattice are given. The exercise of computing the function composition $g(x) = I(f(x))$ over lattice points if f and I are known over lattice points, however, is an easy interpolation problem. One merely computes $f(x)$ for a lattice point x, then locates the four lattice neighbors of $f(x)$ and computes a weighted sum of the values of I over these neighbors. This, so called 'image pullback problem', arises in the texture mapping and morphing. The inverse 'image pushforward' problem is complicated by the fact that the neighbors having the form $f(x)$ of a lattice point in the range of f are not regular. Even computing the set of closest neighbors might be a thorny issue. A solution proposed by Lawton, in his short note *A Fast Algorithm to Map Functions Forward*, uses a duality principle to compute the pushforward indirectly—by first estimating its Fourier coefficients then using an orthonormal scaling function expansion to estimate the pushforward. The pushforward of the (vector valued) identity map on the domain of the function f is then precisely the inverse function.

Acknowledgements

We would again like to thank Prof. N. K. Bose, the Chief editor of the journal for his encouragement, Bob Holland and his editorial staff from Kluwer Academic Publishers for their patience with us, and all the reviewers without whose selfless efforts it would have been impossible to maintain the high quality of the articles. A list of their names appears in the first part of this two part special issues. The first guest co-editor also acknowledges assistance from the National Science Foundation for supporting his work on multidimensional wavelets under NSF contract no. MIP-9696176 to the IBM T. J. Watson research center, Yorktown Heights, New York.

Sankar Basu
IBM T. J. Watson Research Center, Yorktown Heights, NY

Bernard Levy
Dept. of ECE, University of California, Davis, CA

Multidimensional Systems and Signal Processing, 8, 11–30 (1997)

Gröbner Bases and Multidimensional FIR Multirate Systems

HYUNGJU PARK   park@oakland.edu
Department of Mathematical Sciences, Oakland University, Rochester, MI 48309

TON KALKER   kalker@natlab.research.philips.com
Philips Research Laboratories, Prof. Holstlaan 4, 5656 AA, Eindhoven, The Netherlands

MARTIN VETTERLI   martin@eecs.berkeley.edu
Department of Electrical Engineering and Computer Sciences, University of California at Berkeley, Berkeley, CA 94720

Abstract. The **polyphase representation** with respect to sampling lattices in multidimensional (M-D) multirate signal processing allows us to identify perfect reconstruction (PR) filter banks with unimodular Laurent polynomial matrices, and various problems in the design and analysis of invertible MD multirate systems can be algebraically formulated with the aid of this representation. While the resulting algebraic problems can be solved in one dimension (1-D) by the Euclidean Division Algorithm, we show that Gröbner bases offers an effective solution to them in the M-D case.

Key Words: gröbner bases, multirate systems, polyphase representation

1. Introduction

It has been well known that the **polyphase representation** with respect to sampling lattices is a natural representation of multirate systems in studying their algebraic properties. As demonstrated in [1], it allows us to identify various problems in the design and analysis of invertible MD multirate systems with the following mathematical question:

> *Given a matrix of polyphase components, can we effectively decide whether or not that matrix has a left inverse, and give a complete parametrization of all the left inverses of that matrix?*

In this paper, based on the methods initiated in [1], we will further investigate this algebraically simplified problem, and all the systems henceforth will be assumed to be FIR (Finite Impulse Response).

Let us start by reminding the reader that the following three problems were proposed in [1] as demonstrating examples of this theme.

1. Given an MD FIR low-pass filter $G(z)$, decide effectively whether or not $G(z)$ can occur as an analysis filter in a critically downsampled, 2-channel, perfect reconstructing (PR) FIR filter bank. When this decision process yields a positive answer, find **all** such filter banks.

2. Given an oversampled MD FIR analysis filter bank, decide effectively whether or not there is an FIR synthesis filter bank such that the overall system is PR. When this

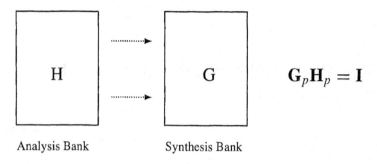

Figure 1. An analysis/synthesis system with PR property.

decision process yields a positive answer, provide a complete parametrization of all such FIR synthesis filter banks.[1]

3. Given a sample rate conversion scheme consisting of upsampling, filtering with an MD FIR filter, and downsampling, decide effectively whether or not this scheme is FIR invertible.

Some of these questions have been studied in 1-D multirate systems (see [3], [4], [5], [2], and [6]), in which the Euclidean Division Algorithm plays a central role. In the M-D case, however, the questions are substantially harder to answer, and it is the goal of this paper to show how Gröbner bases can be used to effectively answer them.

While comprehensive accounts of this theory can be found in [7], [8], [9], [10], [11] and [12], a short review of the Gröbner bases theory is presented in the following section. A heuristic review of this theory can also be found in [13].

2. Gröbner Bases

In order to define Gröbner basis, we first have to introduce the notion of *monomial order*. A monomial in $\mathbb{C}[x] = \mathbb{C}[x_1, \ldots, x_m]$ is a power product of the form $x_1^{e_1} \cdots x_m^{e_m}$, and we denote by $T(x_1, \ldots, x_m)$, or simply by T, the set of all monomials in these variables. In the univariate case, there is a natural monomial order, that is,

$$1 < x < x^2 < x^3 < \cdots.$$

In the multivariate case, we define a monomial order $\leq$ to be a linear order on T satisfying the following two conditions.

1. $1 \leq t$ for all $t \in T$.

2. $t_1 \leq t_2$ implies $t_1 \cdot s \leq t_2 \cdot s$ for all $s, t_1, t_2 \in T$.

Once a monomial order is given, we can talk about the leading monomial (or leading term), $\mathrm{lt}(f(x))$, of $f(x) \in \mathbb{C}[x]$. It should be noted that, if we change the monomial order, then we may have a different $\text{lt}(f(x))$ for the same $f(x)$. Now, fix a monomial order on T, and let $I \subset \mathbb{C}[x]$ be an ideal (i.e a set which includes all the elements which it can generate by taking linear combinations). Define $\text{lt}(I)$ by

$$\text{lt}(I) = \{\text{lt}(f(x)) \mid f \in I\}.$$

DEFINITION 2.1 $\{f_1(x), \ldots, f_t(x\} \subset I$ is called a **Gröbner basis** of I if

$$(\text{lt}(f_1(x)), \ldots, \text{lt}(f_t(x))) = \text{lt}(I)$$

i.e. if the ideal generated by $\text{lt}(f_1(x)), \ldots, \text{lt}(f_t(x)$ coincides with $\text{lt}(I)$.

EXAMPLE 2.2 Fix the degree lexicographic order on $\mathbb{C}[x, y]$, and let I be the ideal generated by $(f(x), g(x))$, with $f(x) = 1 - xy$ and $g(x) = x^2$. Then the relation

$$(1 + xy)f(x) + y^2 g(x) = 1$$

implies that $I = \mathbb{C}[x, y]$, and therefore $\text{lt}(I) = \mathbb{C}[x, y]$. But

$$(\text{lt}(f(x)), \text{lt}(g(x))) = (-xy, x^2) \subset (x).$$

Therefore, $\{f(x), g(x)\}$ is not a Gröbner basis of the ideal I. $\qquad\square$

The main reason that Gröbner basis is useful for us comes from the following analogue of the Euclidean division algorithm.

THEOREM 2.3 (Division Algorithm) *Let* $\{\text{lt}(f_1(x)), \ldots, \text{lt}(f_t(x))\} \subset \mathbb{C}[x]$ *be a Gröbner basis w.r.t. a fixed monomial order, and let* $h(x) \in \mathbb{C}[x]$. *Then there is an algorithm for writing* $h(x)$ *in the form*

$$h(x) = \lambda_1(x)f_1(x) + \cdots + \lambda_t(x)f_t(x) + r(x)$$

such that $h(x) \in I$ *if and only if* $r(x) = 0$.

The polynomial $r(x)$ in the above is called the normal form of $f(x)$ w.r.t. $\{\text{lt}(f_1(x)), \ldots, \text{lt}(f_t(x))\}$. Now, in order to solve our problem, we just compute the normal form of $h(x)$ w.r.t. the given set of polynomials (assuming that this set is a Gröbner basis. Otherwise, we first have to transform it to another set of polynomials which is a Gröbner basis. There is a standard algorithm for this transformation). If it is 0, then $h(x)$ can be written as a linear combination of the polynomials $f_i(x)$ and we have at the same time found the polynomials $\lambda_i(x)$.

REMARK 2.4 There are some results known on the complexity of Gröbner bases computation. If we let

$$\begin{aligned} r &= \text{\# of variables} \\ d &= \text{the maximum degree of the polynomials} \\ s &= \text{the degree of the Hilbert polynomial (this is one less than the} \\ &\quad \text{dimension, and is between 0 and } r - 1) \\ b &= \text{the } \textbf{worst case } \text{upper bound for the degree of the elements of} \\ &\quad \text{the Gröbner basis (of the given polynomials),} \end{aligned}$$

then it is known that

$$b = ((r + 1)(d + 1) + 1)^{(2^{s+1})(r+1)},$$

i.e. is potentially doubly exponential in the number of variables.

This estimate is so large that it seems to suggest that Gröbner bases would be useless in practice. Fortunately, this is not **at all** the case, and the algorithm (in actual use) terminates quite quickly on very many problems of interest. There is a partial understanding of why this is so, and various other bounds are known in some special cases. It is also known that the monomial order being used for the computation **affects** the complexity, i.e. you have to choose a good monomial order in order to shorten the computation time. Reverse degree lexicographic order behaves particularly well in many cases. The papers [14], [15], [16], and [17] contain some results on this complexity issue.

REMARK 2.5 One of the reviewers of this paper asked to include some results on the issue of sensitivity of round-off errors to Gröbner bases computations. We must however note that we are not aware of any results in this direction. The importance of this issue is however recognized by the authors.

3. Unimodularity and Perfect Reconstruction

Since the polyphase representation of an FIR system always gives rise to a rectangular matrix with **Laurent polynomial** entries (see below for a definition), our problem is essentially reduced to the following:

Find left inverses to Laurent polynomial matrices.

DEFINITION 3.1 *Let k be one of the sets* $\mathbb{Q}, \mathbb{R}, \mathbb{C}$.

1. *A* **Laurent polynomial** *f over k in m variables* $x_1, \ldots, x_m$ *is an expression of the form*

$$f(x_1, \ldots, x_m) = \sum_{i_1 = l_1}^{d_1} \cdots \sum_{i_m = l_m}^{d_m} a_{i_1 \cdots i_m} x_1^{i_1} \cdots x_m^{i_m},$$

where $l_1 \le d_1, \ldots, l_m \le d_m$ *are all integers (positive or negative) and each* $a_{i_1 \cdots i_m}$ *is an element of k.*

2. *The set of all the Laurent polynomials over k in $x_1, \ldots, x_m$ is called a* **Laurent polynomial ring**, *and denoted by*

$$k[x^{\pm 1}] := k[x_1^{\pm 1}, \ldots, x_m^{\pm 1}].$$

In order to see why a Laurent polynomial (rather than a polynomial) arises naturally, consider a filter with frequency response

$$H(\omega) = 2\sin(\omega) - 3\cos(2\omega).$$

Then, letting $z := e^{iw}$, we get

$$
\begin{aligned}
H &= 2\frac{e^{iw} - e^{-iw}}{2} - 3\frac{e^{2iw} + e^{-2iw}}{2} \\
&= -\frac{3e^{-2iw}}{2} - e^{-iw} + e^{iw} - \frac{3e^{2iw}}{2} \\
&= -\frac{3}{2z^2} - \frac{1}{z} + z - \frac{3z^2}{2},
\end{aligned}
$$

which is a Laurent polynomial in z.

DEFINITION 3.2 *Let $R := k[x^{\pm 1}]$ be a Laurent polynomial ring.*

1. *Let $\mathbf{v} = (v_1, \ldots, v_n)^t \in R^n$ for some $n \in \mathbb{N}$. Then $\mathbf{v}$ is called a* **unimodular column vector** *if its components generate R, i.e. if there exist $g_1, \ldots, g_n \in R$ such that $v_1 g_1 + \cdots + v_n g_n = 1$.*

2. *A matrix $\mathbf{A} \in M_{pq}(R)$ is called a* **unimodular matrix** *if its maximal minors generate R.*

REMARK 3.3 When $R = \mathbb{C}[x] := \mathbb{C}[x_1, \ldots, x_m]$ is a polynomial ring over $\mathbb{C}$ and the polynomials $v_1, \ldots, v_n \in R$ do not have a common root, Hilbert Nullstellensatz states that there always exist $g_1, \ldots, g_n \in R$ such that $v_1 g_1 + \cdots + v_n g_n = 1$, i.e. $\mathbf{v} = (v_1, \ldots, v_n)^t \in R^n$ is unimodular. In this case, Gröbner bases theory offers a way to find such g_i's (see [7]).

EXAMPLE 3.4 Consider the polynomial matrix $\mathbf{H}_p$ given by

$$
\mathbf{H}_p = \begin{pmatrix} xy - y + 1 & 1 - x \\ yz + w & -z \\ -y & 1 \end{pmatrix} \in M_{32}(k[x, y, z, w]).
$$

Computing (the determinants of) the maximal minors, we find $D_{12}(x) = -w + wx - z$, $D_{13}(x) = -1$ and $D_{23}(x) = w$. Since $0 \cdot D_{12}(x) + (-1)D_{13}(x) + 0 \cdot D_{23}(x) = 1$, $\mathbf{H}_p$ is trivially unimodular. $\qquad\square$

Now the following important result simplifies our problem significantly.

THEOREM 3.5 *A $p \times q$ Laurent polynomial matrix ($p \geq q$) has a left inverse if and only if it is unimodular.*

A proof of this assertion in the case of polynomial matrices can be found in [18], and this result was extended to the case of Laurent polynomial matrices in [19]. An immediate corollary of this theorem is,

COROLLARY 3.6 *An M-D FIR filter bank can be the analysis portion of an M-D PR filter bank if and only if its polyphase matrix is a unimodular Laurent polynomial matrix.*

EXAMPLE 3.7 Consider the polyphase matrix $\mathbf{H}_p$ of the Example 3.4. Since it was shown to be unimodular, it has to have a left inverse. Actually, one verifies easily that

$$\mathbf{G}_p := \begin{pmatrix} 1 & 0 & x - 1 \\ y & 0 & xy - y + 1 \end{pmatrix} \in M_{23}(k[x, y, z, w])$$

satisfies $\mathbf{G}_p \mathbf{H}_p = \mathbf{I}$. This left inverse, however, is far from being unique. On the contrary, a computation using Gröbner bases shows (see [19], page 30–31, page 116) that an arbitrary matrix of the form,

$$\begin{pmatrix} 1 & 0 & x - 1 \\ y & 0 & xy - y + 1 \end{pmatrix} + \begin{pmatrix} u_1 w & -u_1 & u_1(xw - z - w) \\ u_2 w & -u_2 & u_2(xw - z - w) \end{pmatrix}$$

for any Laurent polynomials $u_1, u_2 \in k[x^{\pm 1}, y^{\pm 1}, z^{\pm 1}, w^{\pm 1}]$, is also a left inverse of $\mathbf{H}_p$. Even more strikingly, this parametrization of the left inverses of $\mathbf{H}_p$ in terms of the two parameters u_1 and u_2 turns out to be complete (i.e. exhaustive) and canonical (i.e. minimal with unique representation).

Therefore, the analysis filter bank H whose polyphase matrix is $\mathbf{H}_p$ is FIR invertible, and together with the synthesis filter bank obtained by a backward polyphase superposition of $\mathbf{G}_p$, makes a PR FIR filter bank. And the above parametrization in terms of the two Laurent polynomial parameters u_1, u_2 gives a complete and canonical parametrization of the complementary FIR filter banks. Thus, the **degree of freedom** with which we can design a PR pair of H is precisely 2. □

Therefore, mathematically, we are dealing with the problem of determining if a given **Laurent polynomial** matrix is unimodular, and in case it is, if we can explicitly find all the (not unique in non-square cases) left inverses for it. This allows us to see the study of perfect reconstructing FIR filter banks as the study of unimodular matrices over Laurent polynomial rings.

4. Causal FIR systems and General FIR systems

Many of the known methods for unimodular matrices are developed mainly over polynomial rings, i.e. when the matrices involved are unimodular **polynomial** matrices rather than

Laurent polynomial matrices. In system theoretic terminology, causal invertibility of causal filters are therefore covered by these methods. Geometrically, this demonstrates the relative simplicity associated with affine systems compared to toric systems.

The situation, however, is more complicated partly because an FIR-invertible causal filter may not be causal-invertible. For an example, consider the polynomial vector $\left(\begin{smallmatrix} z \\ z^2 \end{smallmatrix}\right) \in (k[z])^2$. While the relation $\frac{1}{2z} \cdot z + \frac{1}{2z^2} \cdot z^2 = 1$ clearly shows the FIR-invertibility of this vector, it is not causal-invertible since there are no polynomials $f(z), g(z) \in k[z]$ satisfying

$$ f(z) \cdot z + g(z) \cdot z^2 = 1 $$

as we can see easily by evaluating both sides at $z = 0$.

Now, in order to extend any affine results (i.e. causal cases) to general FIR systems, we need an effective process of converting a given Laurent polynomial matrix to a polynomial matrix while preserving unimodularity i.e. we have to perform a preparatory process to convert the problems to causal problems.

We already have presented a systematic method to this effect in [1]: for every variable z_i we introduce two new variables x_i and y_i. Substituting x_i^m for every positive power z_i^m and y_i^k for every negative power z_i^{-k}, we transform the original set of Laurent polynomials into a set of regular polynomials. We then enlarge this set by adding the polynomials $x_i y_i - 1$. One verifies that the constant 1 is a linear combination of the original set of Laurent polynomials if and only if the same is true for the constructed set of regular polynomials. Moreover, given a linear combination of polynomials, we find a linear combination of Laurent polynomials by back substitution: x_i and y_i are replaced by z_i and z_i^{-1} respectively.

There are, however, some drawbacks with this method. First, it significantly increases the complexity of the problem by introducing extra variables and by enlarging the size of the given polynomial vector. Also, a complete parametrization of solutions needs separate computation.

To remedy the situation, an alternative systematic process for the same purpose was developed in [19], and was named the **LaurentToPoly** Algorithm. An input-output description of this algorithm is given in the box of Figure 2. An overview of the main ingredients of this algorithm is presented in the Appendix (see [19] for a complete description of this algorithm).

In this paper, we will mainly rely on this result to reduce the FIR problems to causal FIR problems. A graphical demonstration of this process is shown in the Figure 3.

Now finding an FIR inverse G to the given FIR filter H is equivalent to finding a causal inverse $\hat{G}$ to the causal filter $\hat{H}$.

EXAMPLE 4.1 Consider the unimodular Laurent polynomial vector

$$ \mathbf{v} = (z, z^2)^t \in k[z^{\pm 1}]. $$

As was pointed out in the Section 4, this vector is not unimodular as a **polynomial** vector.

Input:	$\mathbf{v}(x) \in (k[x^{\pm 1}])^n$, a **Laurent** polynomial column vector
Output:	$x \to y$, a change of variables $\mathbf{T}(x) \in \mathrm{GL}_n(k[x^{\pm 1}])$, a square unimodular Laurent polynomial matrix
Specification:	(1) $\hat{\mathbf{v}}(y) := \mathbf{T}(x)\mathbf{v}(x) \in (k[y])^n$ is a **polynomial** column vector in the new variable y (2) $\mathbf{v}(x)$ is unimodular over $k[x^{\pm 1}]$ if and only if $\hat{\mathbf{v}}(y)$ is unimodular over $k[y]$

Figure 2. Algorithm **LaurentToPoly**

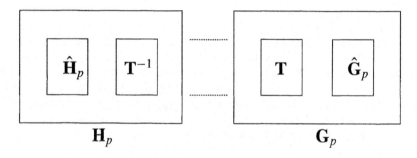

$$\mathbf{H}_p \qquad\qquad\qquad\qquad \mathbf{G}_p$$

Figure 3. Conversion of an FIR system H to a causal system $\hat{H}$

If we let the transformation matrix $\mathbf{T}$ be

$$\mathbf{T} = \begin{pmatrix} 1/z & 0 \\ 0 & z \end{pmatrix},$$

then the converted vector, $\hat{\mathbf{v}} = \mathbf{T}\mathbf{v} = (1, z^3)^t$, is unimodular as a polynomial vector. □

REMARK 4.2 At this point, we would like to make a remark on the opinion expressed in [20]. In that paper, it is stated that the class of 1D *cafacafi* systems is more tractable than the more general class of FIR systems with FIR inverse. The **LaurantToPoly** algorithm, however, shows that this is not necessarily true: using **LaurentToPoly**, any invertible FIR system is translated into the mathematically well understood domain of invertible polynomial matrices. At the same time, **LaurentToPoly** is also applicable in the MD case.

5. 1-D Case

In this section, we will present a complete solution to our problem in the transparent one-dimensional (1-D) setup. And our main tool in this section will be the Euclidean Division Algorithm for the univariate polynomial ring.

5.1. Determination of 1-D FIR Invertibility

Let H be an FIR filter bank whose polyphase matrix is $\mathbf{H}_p \in M_{st}(k[x^{\pm 1}])$, a $s \times t$ univariate Laurent polynomial matrix $(s \geq t)$. Suppose H is FIR invertible. Then, since a Laurent polynomial matrix has a left inverse if and only if it is unimodular, there exist $l := \binom{s}{t}$ Laurent polynomials $D_i(x)$ such that

$$\sum_{i=1}^{l} D_i(x)M_i(x) = 1, \tag{1}$$

where $M_i(x)$ ranges over the maximal minors of $\mathbf{H}_p$.

So, determining the FIR invertibility of H is equivalent to determining the unimodularity of the Laurent polynomial matrix $\mathbf{H}_p$, or equivalently, the unimodularity of the Laurent polynomial vector $(M_1, \ldots, M_l) \in (k[x^{\pm 1}])^l$.

When $k = \mathbb{C}$, this unimodularity determination problem can be readily solved once we notice that, due to the the Laurent polynomial analogue of Hilbert Nullstellensatz over $\mathbb{C}$, $\sum_i D_i(x)M_i(x) = 1$ is possible if and only if the Laurent polynomials $M_i(x)$'s, $1 \leq i \leq \binom{s}{t}$, have no nonzero common roots, i.e. no roots in $\mathbb{C}^*$. Since each univariate Laurent polynomial $M_i(x)$ has only finitely many zeros which can be explicitly found using any existing computer algebra packages, we can tell if $M_i(x)$'s have a nonzero common root or not, and thereby determining if $\mathbf{H}_p$ is unimodular.

EXAMPLE 5.1 Consider a sample rate conversion scheme consisting of upsampling by $p = 3$, filtering with an FIR filter $U(z)$ and downsampling by $q = 2$, where $U(z)$ is given by

$$U(z) = \frac{3}{z^6} + \frac{6}{z^5} + \frac{6}{z^3} + \frac{3}{z^2} - 2 + 29z + 25z^3 + 2z^5 - 2z^6 - 4z^7 + 2z^8$$

$$-23z^9 - 2z^{10} + 4z^{11} + 2z^{12} - 20z^{13} - 16z^{15} + 20z^{17} + 20z^{21}.$$

Then we get the polyphase decomposition $U(z) = \sum_{i=0}^{5} z^i U_i(z^6)$ of $U(z)$ where $U_i(z)$'s are found as

$$U_0(z) = \frac{3}{z} - 2 - 2z + 2z^2$$

$$U_1(z) = \frac{6}{z} + 29 - 4z - 20z^2$$

$$U_2(z) = 2z$$

$$U_3(z) = \frac{6}{z} + 25 - 23z - 16z^2 + 20z^3$$

$$U_4(z) = \frac{3}{z} - 2z$$

$$U_5(z) = 2 + 4z + 20z^2.$$

Now, as demonstrated in [1], the FIR invertibility of the given scheme is equivalent to the FIR invertibility of the following polynomial matrix:

$$\mathbf{U} = \begin{pmatrix} U_0(z) & U_3(z) \\ U_4(z) & U_1(z) \\ U_2(z) & U_5(z) \end{pmatrix}.$$

The three maximal minors of $\mathbf{U}$ are

$$M_1(z) = \begin{vmatrix} U_0(z) & U_3(z) \\ U_4(z) & U_1(z) \end{vmatrix} = -1$$

$$M_2(z) = \begin{vmatrix} U_0(z) & U_3(z) \\ U_2(z) & U_5(z) \end{vmatrix} = \frac{6}{z} - 4 - 2z + 2z^2$$

$$M_3(z) = \begin{vmatrix} U_4(z) & U_1(z) \\ U_2(z) & U_5(z) \end{vmatrix} = \frac{6}{z} - 2z$$

which obviously don't have any common roots.

Consequently the given scheme is FIR invertible.   □

5.2. Parametrization of 1-D PR Pairs

Let the polyphase matrix be $\mathbf{A}$, a $p \times q$ Laurent polynomial matrix, $p \geq q$. Since this polyphase matrix has a left inverse if and only if it is unimodular, we can first determine its unimodularity by the method outlined in the above. If this test shows the unimodularity of $\mathbf{A}$, we first apply the algorithm **LaurentToPoly** to $\mathbf{A}$, converting $\mathbf{A}$ to a unimodular polynomial matrix $\hat{\mathbf{A}}$. Then, by using Euclidean Division Algorithm, we apply a succession of elementary operations to $\hat{\mathbf{A}}$ to reduce it to the following $p \times q$ matrix

$$\begin{pmatrix} \mathbf{I}_q \\ \mathbf{0} \\ \vdots \\ \mathbf{0} \end{pmatrix} \in M_{pq}(k),$$

where $\mathbf{I}_q$ is the $q \times q$ identity matrix, and $\mathbf{0}$ is the q-dimensional zero row vector.

This means that we can find $\mathbf{E} \in E_p(k[z])$, a product of elementary matrices, such that

$$\mathbf{E}\mathbf{A} = \begin{pmatrix} \mathbf{I}_q \\ \mathbf{0} \\ \vdots \\ \mathbf{0} \end{pmatrix}.$$

Now take the first q rows of E to make a $q \times p$ matrix F, i.e.

$$F := (I_q, 0, \ldots, 0)E.$$

Then F is a desired left inverse of A. Note here that $A = E^{-1} \begin{pmatrix} I_q \\ 0 \\ \vdots \\ 0 \end{pmatrix}$ implies $E^{-1} \in$

$GL_p(k[z])$ is a unimodular completion of A.

To get a complete parametrization of all the possible left inverses of A, let $B \in M_{qp}(k[z])$ be an arbitrary left inverse of A. Then

$$BA = I_q.$$

Now, since E^{-1} is a unimodular completion of A,

$$BE^{-1} = (I_q, u_1, \ldots, u_{p-q})$$

for some $u_1, \ldots, u_{p-q} \in (k[z^{\pm 1}])^q$. Now, regarding $u_1, \ldots, u_{p-q}$ as free parameters ranging over q-dimensional Laurent polynomial vectors, we get a complete parametrization of the left inverses to A in terms of $(p-q)q$ parameters ranging over the Laurent polynomials in $k[z^{\pm 1}]$:

$$B = (I_q, u_1, \ldots, u_{p-q})E. \tag{2}$$

REMARK 5.2 If $p = q$, i.e. if the polyphase matrix A is a square unimodular matrix, then the number of free parameters is $(p - q)q = 0$. This coincides with the fact that a square unimodular matrix has a unique inverse.

EXAMPLE 5.3 Consider an oversampled 1-D FIR analysis filter bank whose polyphase matrix is the matrix U of the Example 5.1. We already saw in that example that

$$U = \begin{pmatrix} \frac{3}{z} - 2 - 2z + 2z^2 & \frac{6}{z} + 25 - 23z - 16z^2 + 20z^3 \\ \frac{3}{z} - 2z & \frac{6}{z} + 29 - 4z - 20z^2 \\ 2z & 2 + 4z + 20z^2 \end{pmatrix}$$

is unimodular, so there is an FIR synthesis filter bank such that the overall system is PR. Now we want to find all such FIR synthesis filter banks.

Closely following the algorithm outlined in the above, we get

$$EU = \begin{pmatrix} 1 & 0 \\ 0 & 1 \\ 0 & 0 \end{pmatrix},$$

where the 3×3 matrix E is found as

$$\begin{pmatrix} \frac{z}{18}(-18-125z-188z^2+252z^3-215z^4+178z^5+6z^6) & \frac{z}{3}(-2-27z+30z^2+z^3) & \frac{(-12-89z+51z^2-60z^3-2z^4)}{6} \\ \frac{z}{6}(3+19z-32z^2+23z^3-9z^4-8z^5+6z^6) & z(4-3z-z^2+z^3) & \frac{9}{2}-4z+\frac{3z^2}{2}+z^3-z^4 \\ z(-4z+\frac{23z^2}{3}-5z^3+z^4+\frac{8z^5}{3}-2z^6) & 2z(-3+2z+z^2-z^3) & -6+6z-z^2-2z^3+2z^4 \end{pmatrix}.$$

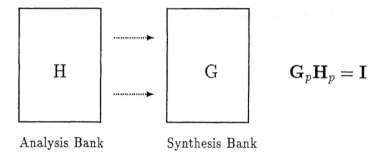

Analysis Bank   Synthesis Bank

Figure 4. Frequency response of the lowpass filter $H(z)$.

Now a general left inverse of $\mathbf{U}$ is in the form

$$\begin{pmatrix} 1 & 0 & u \\ 0 & 1 & v \end{pmatrix} \mathbf{E},$$

where u, v are arbitrary Laurent polynomials in $k[z^{\pm 1}]$.   □

Now we consider a real world problem.

EXAMPLE 5.4   Consider a causal lowpass filter $H(z)$ given by

$$H(z) = 0.1605 + 0.4156z + 0.4592z^2 + 0.1487z^3 - 0.1643z^4 - 0.1245z^5 + 0.0825z^6$$

$$+ 0.0887z^7 - 0.0508z^8 - 0.0608z^9 + 0.0351z^{10} + 0.0399z^{11} - 0.0256z^{12}$$

$$- 0.0244z^{13} + 0.0186z^{14} + 0.0135z^{15} - 0.0131z^{16} - 0.0074z^{17} + 0.0129z^{18} - 0.0050z^{19}$$

whose lowpass characteristic is shown in the Figure 4.

This is decomposed into polyphase components as

$$H(z) = H_0(z^2) + zH_1(z^2),$$

where $H_0(z)$ and $H_1(z)$ are

$$H_0(z) = 0.1605 + 0.4592z - 0.1643z^2 + 0.0825z^3 - 0.0508z^4 + 0.0351z^5 - 0.0256z^6$$

$$+ 0.0186z^7 - 0.0131z^8 + 0.0129z^9,$$

$$H_1(z) = 0.4156 + 0.1487z - 0.1245z^2 + 0.0887z^3 - 0.0608z^4 + 0.0399z^5 - 0.0244z^6$$

$$+ 0.0135z^7 - 0.0074z^8 - 0.0050z^9.$$

the Euclidean Division yields

$$H_0(z) = -2.5893 H_1(z) + r(z)$$

with the remainder

$$r(z) = 1.2367 + 0.8442z - 0.4867z^2 + 0.3123z^3 - 0.208349z^4 + 0.138472z^5 - 0.0888109z^6$$

$$+0.0536797z^7 - 0.0323696z^8.$$

Carrying out the corresponding elementary operation gives

$$\mathbf{E}_{12}(2.5893) \begin{pmatrix} H_0(z) \\ H_1(z) \end{pmatrix} = \begin{pmatrix} r(z) \\ H_1(z) \end{pmatrix}.$$

Repeating the same procedure to the polynomial vector $\begin{pmatrix} r(z) \\ H_1(z) \end{pmatrix}$, we eventually get $\mathbf{C} \in E_2(\mathbb{C}[z])$, a product of 10 elementary matrices, such that

$$\mathbf{C} \begin{pmatrix} H_0(z) \\ H_1(z) \end{pmatrix} = \begin{pmatrix} 0.7661 \\ 0 \end{pmatrix}.$$

Let $\mathbf{E} := \frac{1}{0.7661}\mathbf{C}$. Then

$$\mathbf{E} \begin{pmatrix} H_0(z) \\ H_1(z) \end{pmatrix} = \begin{pmatrix} 1 \\ 0 \end{pmatrix}.$$

An explicit computation shows

$$E_{11}(z) = -0.4138 + 0.5743z - 0.3989z^2 + 0.2652z^3 - 0.1667z^4 + 0.0960z^5 - 0.0478z^6$$

$$+0.0164z^7 + 0.0154z^8$$

$$E_{12}(z) = 2.5658 - 0.6827z + 0.3689z^2 - 0.2369z^3 + 0.1658z^4 - 0.1189z^5 + 0.0839z^6$$

$$-0.0572z^7 + 0.0398z^8$$

$$E_{21}(z) = -0.7081 - 0.2533z + 0.2121z^2 - 0.1512z^3 + 0.1036z^4 - 0.0679z^5 + 0.0416z^6$$

$$-0.0231z^7 + 0.0127z^8 + 0.0085z^9$$

$$E_{22}(z) = 0.2735 + 0.7824z - 0.2799z^2 + 0.1406z^3 - 0.0865z^4 + 0.0599z^5 - 0.0436z^6$$

$$+0.0317z^7 - 0.0223z^8 + 0.0220z^9.$$

Now by the parametrization formula (2), any left inverse $\mathbf{B}$ is of the form

$$\mathbf{B}(u) = (1 \quad u)\mathbf{E}$$

$$= (E_{11} + uE_{21} \quad E_{12} + uE_{22})$$

$$= (E_{11} \quad E_{12}) + u(E_{21} \quad E_{22}),$$

where $u \in k[z^{\pm 1}]$ is an arbitrary Laurent polynomial. Now, the 1 parameter family of filters

$$F(z, u(z)) = B_{11}(z^2, u(z^2)) + zB_{12}(z^2, u(z^2)), \quad u \in k[z^{\pm 1}]$$

describes all the synthesis filters, and making a good choice of $u \in k[z^{\pm 1}]$ will give us a synthesis filter with a more desirable frequency response. $\qquad \square$

6. Gröbner Bases and M-D FIR Systems

Let an $s \times t$ matrix $\mathbf{H}_p \in \mathrm{M}_{st}(k[x^{\pm 1}])$ be the polyphase matrix of a given M-D FIR filter bank H ($s \geq t$). Now, suppose we want to find a synthesis filter bank G so that G together with H makes a perfect reconstructing system.

Applying the **LaurentToPoly** Algorithm to $\mathbf{H}_p \in \mathrm{M}_{ts}(k[x^{\pm 1}])$ to obtain $\hat{\mathbf{H}}_p \in \mathrm{M}_{ts}(k[x])$, we see that this problem is equivalent to finding a $t \times s$ matrix $\hat{\mathbf{G}}_p \in \mathrm{M}_{ts}(k[x])$ such that $\hat{\mathbf{G}}_p \hat{\mathbf{H}}_p = \mathbf{I}_q$. After getting such a $\hat{\mathbf{G}}_p$, we can apply the **LaurentToPoly** Algorithm *backwards* to $\hat{\mathbf{G}}_p$ to obtain $\mathbf{G}_p \in \mathrm{M}_{ts}(k[x^{\pm 1}])$. Then it follows that $\mathbf{G}_p \mathbf{H}_p = \mathbf{I}_q$, and the filter bank G with its polyphase matrix being $\mathbf{G}_p$ is a desired synthesis filter bank.

So, we have reduced our problem to

> For a given **polynomial matrix** $\mathbf{A} \in \mathrm{M}_{st}(k[x])$ ($s \geq t$), find a (particular) left inverse $\mathbf{B} \in \mathrm{M}_{ts}(k[x])$ of $\mathbf{A}$.

This is in fact possible using Gröbner bases, and the following method was introduced in [18], and was exploited in our context in [19]:

The column vectors of the unimodular matrix $\mathbf{A}^t = (f_{ji}) \in \mathrm{M}_{st}(k[x])$ span the free $k[x]$-module $(k[x])^q$. Therefore, we can use Gröbner bases to express the standard basis vectors $\mathbf{e}_1, \dots \mathbf{e}_q \in (k[x])^q$ as linear combinations (with polynomial coefficients) of the column vectors of $\mathbf{A}^t$.

More explicitly, denoting the i-th column vector of $\mathbf{A}^t$ by $\mathbf{w}_i$, $1 \leq i \leq p$, we have $\mathbf{w}_i := \begin{pmatrix} f_{i1} \\ \vdots \\ f_{iq} \end{pmatrix}$. Now, use Gröbner bases to find g_{ij}'s such that

$$\mathbf{e}_1 := \begin{pmatrix} 1 \\ 0 \\ \vdots \\ 0 \end{pmatrix} = g_{11}\mathbf{w}_1 + \cdots + g_{1p}\mathbf{w}_p = g_{11}\begin{pmatrix} f_{11} \\ \vdots \\ f_{1q} \end{pmatrix} + \cdots + g_{1p}\begin{pmatrix} f_{p1} \\ \vdots \\ f_{pq} \end{pmatrix}$$

$$\vdots$$

$$\mathbf{e}_q := \begin{pmatrix} 0 \\ \vdots \\ 0 \\ 1 \end{pmatrix} = g_{q1}\mathbf{w}_1 + \cdots + g_{qp}\mathbf{w}_p = g_{q1}\begin{pmatrix} f_{11} \\ \vdots \\ f_{1q} \end{pmatrix} + \cdots + g_{qp}\begin{pmatrix} f_{p1} \\ \vdots \\ f_{pq} \end{pmatrix}.$$

Denoting the $q \times p$ matrix (g_{ij}) by $\mathbf{B}$, we can rewrite the above set of equations as

$$\mathbf{I}_q = \begin{pmatrix} g_{11} & \cdots & g_{1p} \\ \vdots & & \vdots \\ g_{q1} & \cdots & g_{qp} \end{pmatrix}\begin{pmatrix} f_{11} & \cdots & f_{1q} \\ \vdots & & \vdots \\ f_{p1} & \cdots & f_{pq} \end{pmatrix} = \mathbf{BA}.$$

And this $\mathbf{B}$ is precisely (one of) what we want.

EXAMPLE 6.1 Again, consider the polyphase matrix

$$\mathbf{H}_p = \begin{pmatrix} xy - y + 1 & 1 - x \\ yz + w & -z \\ -y & 1 \end{pmatrix} \in M_{32}(k[x, y])$$

of Example 3.4. The following is a SINGULAR[2] script implementing the algorithm of this section to find a $\mathbf{G}_p \in M_{23}(k[x, y])$ such that $\mathbf{G}_p\mathbf{H}_p = \mathbf{I}_2$.

```
ring r=0,(x,y,z,w),(c,dp);option(redSB);
vector v(1)=[xy-y+1,1-x];vector v(2)=[yz+w,-z];
vector v(3)=[-y,1];
module M=v(1),v(2),v(3);
module G=std(M); matrix T=lift(M,G);
```

And the results are

```
> G;
G[1]=[0,1]
G[2]=[1]

> T;
T[1,1]=y
T[1,2]=1
T[2,1]=0
T[2,2]=0
T[3,1]=xy-1y+1
T[3,2]=x-1
```

Since $\{(1, 0), (0, 1)\}$ is a Gröbner basis of the row vectors of $\mathbf{H}_p$, $\mathbf{H}_p$ is unimodular, and the relation $\mathtt{G} = \mathtt{MT}$ translates to

$$\begin{pmatrix} 0 & 1 \\ 1 & 0 \end{pmatrix} = \mathbf{H}_p^t \mathtt{T}.$$

By taking transpose of both sides, we get $\mathtt{T}^t\mathbf{H}_p = \begin{pmatrix} 0 & 1 \\ 1 & 0 \end{pmatrix}$, i.e.

$$\begin{pmatrix} 1 & 0 & x - 1 \\ y & 0 & xy - y + 1 \end{pmatrix} \mathbf{H}_p = \mathbf{I}_2.$$

Hence, $\mathbf{G}_p := \begin{pmatrix} 1 & 0 & x - 1 \\ y & 0 & xy - y + 1 \end{pmatrix}$ is a left inverse of $\mathbf{H}_p$. $\square$

EXAMPLE 6.2 Consider the 2D sample rate conversion scheme which consists of vertical upsampling by a factor 3, filtering with a filter $H(z) = H(z_1, z_2)$ and horizontal downsampling with a factor 2. We assume that H is FIR, and we would like to know if this scheme has an FIR inverse. To be more precise, we are looking for an FIR filter $G(z)$, such that horizontal upsampling by a factor 2, filtering with $G(z)$ and vertical downsampling with a factor 3, cancels the effect of the first sample rate conversion scheme.

Let the filter $H(z)$ be given by $H(z) = \sum h_{i,j} z_1^i z_2^j$. Following the method outlined in Section 5, but now for this 2D case, we construct the 3×2 polynomial matrix $H_{k,l}(z) = \sum h_{3i+k, 2j+l} z_1^i z_2^j$, where $0 \le k \le 2$ and $0 \le l \le 1$.

Assume momentarily that $H(z)$ is a separable filter $H^h(z_1) H^v(z_2)$. It is easily seen that in this case the filters $H_{k,l}(z)$ are products of 1D polyphase components, i.e. $H_{k,l}(z) = H_k^h(z_1) H_l^v(z_2)$. Consequently, all the maximal minors of $H_{k,l}(z)$ have determinants equal to 0. Therefore the 2D analogue of Eq. 1 cannot be satisfied, and inversion is impossible.

Now we consider a non-separable case, where the filter $H(z)$ is given by the 4×6 (horizontal $\times$ vertical) impulse response

$$\begin{pmatrix} 2 & 3 & 2 & 1 & 3 & 2 \\ 3 & 5 & 3 & 1 & 3 & 2 \\ 1 & 1 & 1 & 1 & 1 & 1 \\ 2 & 2 & 2 & 1 & 1 & 1 \end{pmatrix}.$$

The polyphase component matrix $\mathbf{H}_p = (H_{kl})$ of $H(z)$ is defined by

$$H(z) = \sum_{k=0}^{2} \sum_{l=0}^{1} z_1^k z_2^l H_{kl}(z_1^2, z_2^3),$$

which is found as

$$\mathbf{H}_p = \begin{pmatrix} 2 + z_1 + z_2 + z_1 z_2 & 3 + 2z_1 + z_2 + z_1 z_2 \\ 3 + z_1 + 3z_2 + z_1 z_2 & 5 + 2z_1 + 3z_2 + z_1 z_2 \\ 2 + z_1 + 2z_2 + z_1 z_2 & 3 + 2z_1 + 2z_2 + z_1 z_2 \end{pmatrix}.$$

Computing the determinants of the maximal minors we find $D_0(z) = -1 - z_2$, $D_1(z) = -z_2 - z_1 z_2$ and $D_2(z) = 1 - z_2 - z_1 z_2$. These determinants are proper multivariable expressions and the Euclidean algorithm will therefore not work. In this case one easily verifies that $D_2 - D_1 = 1$, and therefore $\mathbf{H}_p$ is unimodular and there exist an inverse FIR filter $G(z)$. To find $G(z)$ we first need to find a left inverse $\mathbf{G}_p$ to $\mathbf{H}_p$.

The following is an example SINGULAR script used to compute a left inverse of $\mathbf{H}_p$. For notational convenience, we let $x := z_1$, $y := z_2$, $\mathbf{A} := \mathbf{H}_p$, and $\mathbf{B} := \mathbf{G}_p$.

```
ring r=0,(x,y),(c,dp); option(redSB);
vector v(1)=[2+x+y+xy,3+2*x+y+xy];
vector v(2)=[3+x+3*y+xy,5+2*x+3*y+xy];
vector v(3)=[2+x+2*y+xy,3+2*x+2*y+xy];
module M=v(1),v(2),v(3);
module G=std(M); matrix T=lift(M,G)
```

The output from SINGULAR is as follows:

```
> G;
G[1]=[0,1]
G[2]=[1]
```

```
> T;
T[1,1]=0
T[1,2]=1
T[2,1]=x+2
T[2,2]=-2x-3
T[3,1]=-1x-3
T[3,2]=2x+4
```

Since $\{(1, 0), (0, 1)\}$ is a Gröbner basis of the row vectors of $\mathbf{A}$, $\mathbf{A}$ is unimodular, and the relation $G = MT$ translates to

$$\begin{pmatrix} 0 & 1 \\ 1 & 0 \end{pmatrix} = \mathbf{A}'T.$$

By taking transpose of both sides, we get $T'\mathbf{A} = \begin{pmatrix} 0 & 1 \\ 1 & 0 \end{pmatrix}$, i.e.

$$\begin{pmatrix} 1 & -2x-3 & 2x+4 \\ 0 & x+2 & -x-3 \end{pmatrix} \mathbf{A} = \mathbf{I}_2.$$

Hence, $\mathbf{B} := \begin{pmatrix} 1 & -2z_1-3 & 2z_1+4 \\ 0 & z_1+2 & -z_1-3 \end{pmatrix}$ is a left inverse of $\mathbf{A}$. $\square$

Appendix

A. LaurentToPoly Algorithm

In this appendix, we present an algorithm that transforms a Laurent polynomial column vector to a polynomial column vector while preserving unimodularity. A schematic description of this **LaurentToPoly** Algorithm was given in Section 4, and we refer to the Figure 2 for the notations. This process is very powerful essentially because the unimodularity of the Laurent polynomial vector $\mathbf{v}(x) \in (k[x^{\pm 1}])^n$ is converted to the unimodularity of the polynomial vector $\hat{\mathbf{v}}(y) \in (k[x])^n$. For the results stated in this appendix without proof, see [19].

We start with a theorem (without proof) that can be seen as an analogue of the Noether Normalization Lemma. The Noether Normalization Lemma states that, for any given polynomial $f \in k[x]$, by defining new variables $y_1, \ldots, y_m$ by $x_1 = y_1, x_2 = y_2 + y_1^l, \ldots, x_m = y_m + y_1^{l^{m-1}}$ for a sufficiently large $l \in \mathbb{N}$ and regarding f as a polynomial in the new variables $y_1, \ldots, y_m$, we can make f a monic polynomial in the first variable y_1. Now, we extend this to the Laurent polynomial ring $k[x^{\pm 1}] = k[x_1^{\pm 1}, \ldots, x_m^{\pm 1}]$.

THEOREM A.1 (Laurent polynomial analogue of Noether Normalization) *Let $f \in k[x^{\pm 1}]$ be a Laurent polynomial, and define new variables $y_1, \ldots, y_m$ by $x_1 = y_1, x_2 = y_2 y_1^l, \ldots, x_m = y_m y_1^{l^{m-1}}$. Then, for a sufficiently large $l \in \mathbb{N}$, the leading and the lowest coefficients of $f \in k[x^{\pm 1}]$ with respect to the first variable y_1 are units in the ring $k[y_2^{\pm 1}, \ldots, y_m^{\pm 1}]$.*

Input:	$f \in k[x^{\pm 1}]$
Output:	$x \to y$, a change of variables
Specification:	the leading and the lowest coefficients of $f \in k[y^{\pm 1}]$ with respect to the first variable y_1 are units in the ring $k[y_2^{\pm 1}, \ldots, y_m^{\pm 1}]$

Figure 5. **LaurentNoether**

With this theorem at hand, we can now describe the **LaurentToPoly** Algorithm.

Let $n \geq 2$, $S = k[x_2^{\pm 1}, \ldots, x_m^{\pm 1}]$, and $\mathbf{v} = (v_1, \ldots, v_n)^t \in (k[x^{\pm 1}])^n = (S[x_1])^n$. By using the algorithm **LaurentNoether**, we may assume that the leading and the lowest coefficients of v_1 w.r.t. x_1 are invertible elements of S. Write

$$v_1 = a_p x_1^p + a_{p+1} x_1^{p+1} + \cdots + a_q x_1^q$$

where a_p and a_q are units of S.

- **Step 1:** Using the invertibility of $a_p \in S$, define $\mathbf{D} \in M_n(S[x_1^{\pm 1}])$ and $\mathbf{v}' \in (S[x_1^{\pm 1}])^n$ by

$$\mathbf{D} := \begin{pmatrix} a_p^{-1} x_1^{-p} & 0 & \\ 0 & a_p x_1^p & \\ & & I_{n-2} \end{pmatrix}$$

$$\mathbf{v}' = (v_1', \ldots, v_n')^t := \mathbf{D}\mathbf{v}.$$

Note here that the matrix

$$\mathbf{D} = E_{21}(a_p x_1^p) E_{12}(1 - a_p^{-1} x_1^{-p}) E_{21}(1) E_{12}(1 - a_p x_1^p)$$

is realizable over $S[x_1^{\pm 1}]$, and

$$v_1' = a_p^{-1} x_1^{-p} v_1 = 1 + a_{p+1}/a_p x_1 + \cdots + a_q/a_p x_1^{q-p}$$

is a **polynomial** in $S[x_1]$.

- **Step 2:** Since the constant term of $v_1' \in S[x_1]$ is 1, by adding suitable multiples of v_1' to v_i''s, $i = 2, \ldots, n$, we can make $v_2', \ldots, v_n'$ polynomials in $S[x_1]$ whose constant terms are zero, i.e. find $E \in E_n(k[x^{\pm 1}])$ such that

$$E\mathbf{v}' = \hat{\mathbf{v}} = \begin{pmatrix} \hat{v}_1 \\ \vdots \\ \hat{v}_n \end{pmatrix} \in (S[x_1])^n,$$

where $\hat{v}_1 \equiv 1 \bmod x_1$ and $\hat{v}_i \equiv 0 \bmod x_1$ for all $i = 2, \ldots, n$.

- **Step 3:** Choose a sufficiently large number $l \in \mathbb{N}$ so that, with the following change of variables,

$$x_1 = y_1 \cdot (y_2 \cdots y_m)^l$$
$$x_2 = y_2$$
$$\vdots$$
$$x_m = y_m,$$

all the $\hat{v}_i$'s become polynomials in $k[y]$. Then $\hat{v}_1 \equiv 1 \bmod y_1 \cdots y_m$.

Now give the transformation matrix $\mathbf{T} := \mathbf{ED}$ as the output.

$\square$

It still remains to show that the outcome of this algorithm is what we want. This, however, follows from the following theorem (without proof).

THEOREM A.2 *With the notations as in the above,* $\mathbf{v}(x)$ *is unimodular over* $k[x^{\pm 1}]$ *if and only if* $\hat{\mathbf{v}}(y)$ *is unimodular over* $k[y]$.

Notes

1. This is often called the **complementary filter** problem [2].
2. SINGULAR is a computer algebra system for singularity theory and algebraic geometry, developed in the University of Kaiserslautern, Germany. It is being alpha-tested, and is freely available by anonymous ftp. For more information, see [21].

References

1. T. Kalker, H. Park, and M. Vetterli, "Groebner Bases Techniques in Multidimensional Multirate Systems," *Proceedings of ICASSP 95*, 1995.
2. M. Vetterli and J. Kovačević, *Wavelets and Subband Coding*, Prentice Hall Signal Processing Series: Prentice Hall, 1995.
3. M. Vetterli, "Filter banks allowing perfect reconstruction," *IEEE Transactions on Signal Processing*, vol. 10, 1986, pp. 219–244.
4. A. J. E. M. Janssen, "Note on a linear system occurring in perfect reconstruction," *IEEE Transactions on Signal Processing*, vol. 18, no. 1, 1989, pp. 109–114.
5. M. Vetterli and C. Herley, "Wavelets and filter banks: Theory and design," *IEEE Transactions on ASSP*, vol. 40, 1992, pp. 2207–2232.
6. P. P. Vaidyanathan, *Multirate Systems and Filter Banks*, Prentice Hall Signal Processing Series: Prentice Hall, 1993.
7. B. Buchberger, "Gröbner bases—an algorithmic method in polynomial ideal theory," in *Multidimensional systems theory* (N. K. Bose, ed.), 1985, pp. 184–232. Dordrecht: D. Reidel.
8. D. Cox, J. Little, and D. O'Shea, *Ideals, Varieties, and Algorithms*, Undergraduate Texts in Mathematics: Springer-Verlag, 1992.
9. B. Mishra, *Algorithmic Algebra*, Texts and Monographs in Computer Science: Springer-Verlag, 1993.

10. W. Adams and P. Loustaunau, *An Introduction to Gröbner Bases*, vol. 3 of *Graduate Studies in Mathematics*. American Mathematical Society, 1994.

11. T. Becker and V. Weispfenning, *Gröbner Bases*, vol. 141 of *Graduate Texts in Mathematics*. Springer-Verlag, 1993.

12. D. Eisenbud, *Introduction to Commutative Algebra with a View Towards Algebraic Geometry*, Graduate Texts in Mathematics: Springer-Verlag, 1995.

13. X. Chen, I. Reed, T. Helleseth, and T. K. Truong, "Use of Gröbner bases to Decode Binary Cyclic Codes up to the True Minimum Distance," *IEEE Transactions on Information Theory*, vol. 40, 1994, pp. 1654–1660.

14. D. Bayer and M. Stillman, "A theorem on redefining division orders by the reverse lexicographic orders," *Duke J. of Math.*, vol. 55, 1987, pp. 321–328.

15. D. Bayer and M. Stillman, "On the complexity of computing syzygies," *J. Symb. Comp.*, vol. 6, 1988, pp. 135–147.

16. Gruson, Lazarsfeld, and Peskine, "On a theorem of Castelnuovo, and the equations defining space curves," *Invent. Math.*, vol. 72, 1983, pp. 491–506.

17. Winkler, "On the complexity of the Gröbner basis algorithm over $k[x, y, z]$," in *Springer Lect. Notes in Computer Sci.* 174 (J. Fitch, ed.), Springer-Verlag, 1984.

18. A. Logar and B. Sturmfels, "Algorithms for the Quillen-Suslin theorem," *Journal of Algebra*, vol. 145, 1992, pp. 231–239.

19. H. Park, *A Computational Theory of Laurent Polynomial Rings and Multidimensional FIR Systems*. PhD thesis, University of California at Berkeley, 1995.

20. P. P. Vaidyanathan and T. Chen, "Role of anticausal inverses in multirate filter banks—part i: System-theoretic fundamentals," *IEEE Transactions on Signal processing*, vol. 43, 1995, pp. 1090–1102.

21. G. Greuel, G. Pfister, and H. Schoenemann, *Singular User Manual*, University of Kaiserslautern, version 0.90 ed., 1995. (Available by anonymous ftp from helios.mathematik.uni-kl.de.)

22. H. Park and C. Woodburn, "An Algorithmic Proof of Suslin's Stability Theorem for Polynomial Rings," *Journal of Algebra*, vol. 178, pp. 277–298, 1995.

23. S. Basu and H. M. Choi, "Multidimensional causal, stable, perfect reconstruction filter banks," *Proceedings of the ICIP-94*, vol. 1, 1994, pp. 805–809.

24. Z. Cvetković and M. Vetterli, "Oversampled Filtered Banks," *IEEE Transactions on Signal Processing*, submitted.

Multidimensional Systems and Signal Processing, 8, 31–69 (1997)

Reconstruction and Decomposition Algorithms for Biorthogonal Multiwavelets

CHARLES A. MICCHELLI
IBM T. J. Watson Research Center, P.O. Box 218, Yorktown Heights, New York 10598

YUESHENG XU
Department of Mathematics, North Dakota State University, Fargo, ND 58105

Abstract. We construct biorthogonal multiwavelets (abbreviated to wavelets) in a weighted Hilbert space $L^2(E, \rho)$ where E is a compact subset in $\mathbb{R}^d$. A recursive formula for biorthogonal wavelet construction is presented. The construction of the initial wavelets is reformulated as the solution of a certain matrix completion problem. A general solution of the matrix completion problem is identified and expressed conveniently in terms of any given particular solution. Several particular solutions are proposed. Reconstruction and decomposition algorithms are developed for the biorthogonal wavelets. Special results for the univariate case $E = [0, 1]$ are obtained.

Key Words: biorthogonal multiwavelets, refinable vector fields, reconstruction and decomposition algorithms

1. Introduction

This paper continues the theme of our two recent papers [1, 2] on the construction of discontinuous wavelets. It is known that such wavelets are very useful for solutions of integral equations because multiscale methods based on these wavelets lead to linear systems with sparse coefficient matrices which have bounded condition numbers [3, 4, 5, 6, 7, 8, 9, 10, 11, 12]. To extend upon the applicability of these methods, we construct biorthogonal wavelets in a weighted Hilbert space $L^2(E, \rho)$ where E is a compact subset in $\mathbb{R}^d$. These wavelets are potentially useful in numerical solutions of boundary integral equations defined on a curve or a surface, from which this work is partially motivated.

This paper is organized as follows: In Sections 2 to 4, we present a general framework for the construction of biorthogonal wavelets in a weighted Hilbert space, where a recursive formula is developed which allows for a simple construction of biorthogonal wavelets from one level to another. In Section 5, we construct the initial wavelets by reformulating the problem as the solution of certain matrix equations for which a general solution is found. In Section 6, we develop reconstruction and decomposition algorithms for biorthogonal wavelets. In Section 7, we briefly discuss the construction of certain measure preserving transformations used in Section 2 and special emphasis is put on the case $E = [0, 1]$.

2. Mutually Orthogonal Isometries

Suppose that ρ is a probability measure defined on a compact subset E of $\mathbb{R}^d$ which has no atoms and is absolutely continuous with respect to the Lebesgue measure. We denote by

$L^2(E, \rho)$ the space of the square integrable functions with respect to ρ, i.e.,

$$\|f\| = \|f\|_{L^2(E,\rho)} := \left(\int_E |f(x)|^2 d\rho(x) \right)^{\frac{1}{2}} < \infty, \quad f \in L^2(E, \rho), \tag{2.1}$$

which is a Hilbert space with the inner product

$$\langle f, g \rangle = \langle f, g \rangle_{L^2(E,\rho)} := \int_E f(x)g(x)d\rho(x). \tag{2.2}$$

In this section we establish a general framework for the construction of biorthogonal wavelets in the Hilbert space $L^2(E, \rho)$. For this purpose, we decompose E into a finite number of measurable subsets $E_i, i = 0, 1, \ldots, m - 1$, such that

$$E = \bigcup_{i=0}^{m-1} E_i$$

and

$$\text{meas}\,(E_i \cap E_j) = 0, \quad i \neq j, \quad i, j = 0, 1, \ldots, m - 1,$$

where meas(A) denotes the Lebesgue measure of a set $A \subseteq \mathbb{R}^d$.

Central to our construction are m bijective maps

$$\phi_i : E \to E_i, \quad i = 0, 1, \ldots, m - 1,$$

defined a.e., (which depend on ρ) such that the bounded linear operators defined by

$$(P_i f)(x) = \begin{cases} \frac{1}{\sqrt{\gamma_i}} f(\phi_i^{-1}(x)), & x \in E_i, \\ 0, & x \notin E_i, \end{cases} \quad i = 0, 1, \ldots, m - 1, \tag{2.3}$$

with

$$\gamma_i := \int_{E_i} d\rho(x), \quad i = 0, 1, \ldots, m - 1$$

satisfy the equation

$$\frac{1}{\gamma_i} \int_{E_i} f(\phi_i^{-1}(x))d\rho(x) = \int_E f(x)d\rho(x), \quad f \in L^1(E, \rho). \tag{2.4}$$

That is, each P_i is an *isometry* on $L^2(E, \rho)$. The existence of these mappings follows from general measure theoretic considerations, cf. [13, Theorem 9, p. 270].

Let us recall here an important special case considered in [1]. Assume that E is a compact subset of $\mathbb{R}^d$ which has the property that there are m contractive affine maps $\psi_i : \mathbb{R}^d \to \mathbb{R}^d$,

$i = 0, 1, \ldots, m - 1$, for which the absolute value of their Jacobians are all equal to $\frac{1}{m}$, viz.,

$$|J_{\psi_i}| = \frac{1}{m}, \quad i = 0, 1, \ldots, m - 1,$$

$$E = \bigcup_{i=0}^{m-1} \psi_i(E), \tag{2.5}$$

and

$$\text{meas}(\psi_i(E) \cap \psi_j(E)) = 0, \quad i \neq j, \ i, j = 0, 1, \ldots, m - 1. \tag{2.6}$$

In this case, we choose $E_i = \psi_i(E)$ and $\phi_i = \psi_i$, $i = 0, 1, \ldots, m - 1$.

For example, let M be a $d \times d$ integer matrix such that $\rho(M^{-1}) :=$ spectral radius of $M^{-1} < 1$. Let $m := |\det M|$ and recall that the number of disjoint subsets of $\mathbb{Z}^d / M\mathbb{Z}^d$ is m. Suppose that $\mathcal{M} = \{r_1, \ldots, r_m\}$ is any set of m lattice points of $\mathbb{Z}^d$ that represent distinct cosets of $\mathbb{Z}^d / M\mathbb{Z}^d$. According to [14], corresponding to the contractive (relative to some norm on $\mathbb{R}^d$) affine maps

$$\psi_i(x) := M^{-1}(x + r_i), \quad i = 1, 2, \ldots, m,$$

there is a compact set E which satisfies (2.5), see also [15]. Therefore as long as the set $\mathcal{M}$ can be chosen so that (2.6) holds the maps above provide a set E of the type that we consider. For some concrete examples of this circumstance see [15, 1].

Returning to the general case, we now define linear operators G_i on $L^2(E, \rho)$ by setting

$$G_i f := \sqrt{\gamma_i} f \circ \phi_i, \quad i = 0, 1, \ldots, m - 1, \ f \in L^2(E, \rho). \tag{2.7}$$

Here we are using the notation $(f \circ g)(x) := f(g(x))$, $x \in E$ for function composition. It can be verified that

$$P_i G_i = \chi_{E_i} I, \quad i = 0, 1, \ldots, m - 1 \tag{2.8}$$

and

$$G_i P_i = I, \quad i = 0, 1, \ldots, m - 1 \tag{2.9}$$

where I denotes the identity operator in $L^2(E, \rho)$ and χ_{E_i} denotes the characteristic function of set E_i. Moreover, using equation (2.4) we see that the adjoint of P_i is given by

$$P_i^* = G_i, \quad i = 0, 1, \ldots, m - 1 \tag{2.10}$$

and there holds the formula

$$P_i^* P_j = \delta_{ij} I, \quad i, j = 0, 1, \ldots, m - 1. \tag{2.11}$$

We now choose an $m \times m$ orthogonal matrix

$$Q = (q_{ij})_{i,j=0,1,\ldots,m-1},$$

that is,

$$QQ^T = Q^TQ = I, \tag{2.12}$$

and define a set of linear operators T_i by

$$T_i = \sum_{j=0}^{m-1} q_{ij} P_j, \quad i = 0, 1, \ldots, m-1. \tag{2.13}$$

Equivalently, we have for $g \in L^2(E, \rho)$,

$$(T_i g)(t) = \sum_{j=0}^{m-1} q_{ij} \frac{1}{\sqrt{\gamma_j}} g(\phi_j^{-1}(t)) \chi_{E_j}(t), \quad t \in E, \quad i = 0, 1, \ldots, m-1.$$

Hence, it follows from equations (2.10) and (2.13) that the adjoint of T_i can be written as

$$T_i^* = \sum_{j=0}^{m-1} q_{ij} G_j. \tag{2.14}$$

The following lemma lists some properties of the linear operators $T_i, i = 0, 1, \ldots, m-1$, which are needed later.

LEMMA 2.1 *The following statements hold:*
(i) $T_i^* T_j = \delta_{ij} I$, $i, j = 0, 1, \ldots, m-1$,
(ii) $\langle f, g \rangle = \sum_{i=0}^{m-1} \langle G_i f, G_i g \rangle$, $f, g \in L^2(E, \rho)$.

Proof: (i) To prove this identity, we use equation (2.11). Thus, for $i, j = 0, 1, \ldots, m-1$, we have

$$\begin{aligned}
T_i^* T_j &= \sum_{\ell=0}^{m-1} q_{i\ell} \sum_{\ell'=0}^{m-1} q_{j\ell'} P_\ell^* P_{\ell'} \\
&= \sum_{\ell=0}^{m-1} q_{i\ell} q_{j\ell} I \\
&= \delta_{ij} I.
\end{aligned}$$

(ii) According to (2.8) and (2.10), we have

$$\langle f, g \rangle = \left\langle \sum_{i=0}^{m-1} P_i G_i f, g \right\rangle = \sum_{i=0}^{m-1} \langle G_i f, G_i g \rangle. \qquad \blacksquare$$

There is another sequence of linear operators which will be useful for our wavelet construction. To this end we define for $i = 0, 1, \ldots, m - 1$ linear operators on $L^2(E, \rho)$ by the formula

$$H_i = \sum_{j=0}^{m-1} q_{ji} T_j. \tag{2.15}$$

LEMMA 2.2 *The linear operators H_i defined by (2.15) satisfy the identity*

$$\sum_{i=0}^{m-1} H_i G_i = I.$$

Proof: Let $f \in L^2(E, \rho)$. By the definition of H_i, we have

$$g := \sum_{i=0}^{m-1} H_i G_i f = \sum_{i=0}^{m-1} \sum_{j=0}^{m-1} q_{ji} T_j G_i f.$$

However, equation (2.13) implies that

$$g = \sum_{i=0}^{m-1} \sum_{\ell=0}^{m-1} \left(\sum_{j=0}^{m-1} q_{ji} q_{j\ell} \right) P_\ell G_i f.$$

By (2.12), we have

$$\sum_{j=0}^{m-1} q_{ji} q_{j\ell} = \delta_{i\ell},$$

and so it follows from the above equation and (2.8) that

$$g = \sum_{i=0}^{m-1} P_i G_i f = \sum_{i=0}^{m-1} \chi_{E_i} f = f.$$

This completes the proof. ∎

The linear operators $T_i, i = 0, 1, \ldots, m - 1$, will allow us to generate *iteratively* biorthogonal bases in $L^2(E, \rho)$. We call these bases *biorthogonal wavelet bases* because they are generated *recursively* from the finite number of linear operators $T_0, \ldots, T_{m-1}$. The initial finite dimensional subspace in our recursive procedure must be chosen with care. It is this issue that we describe next.

3. Refinable Vector Fields

Given mappings

$$\phi_i : E \to E_i, \quad i = 0, 1, \ldots, m-1$$

and corresponding linear operators $G_i, i = 0, 1, \ldots, m-1$ on $L^2(E, \rho)$ defined by (2.7) we say $\mathbf{f} := (f_1, \ldots, f_n)^T$ is a *refinable vector field* if there are $n \times n$ matrices $A_0, \ldots, A_{m-1}$ such that

$$G_i \mathbf{f} = \sqrt{\gamma_i} A_i^T \mathbf{f}, \quad i = 0, 1, \ldots, m-1 \tag{3.1}$$

and

$$\mathbf{v}^T \mathbf{f} = a \tag{3.2}$$

where $\mathbf{v} \in \mathbb{R}^n$ and a is some nonzero constant.

As a matter of notation we always use the convention that any bounded linear operator T on $L^2(E, \rho)$ acts componentwise on vector fields. That is, when $\mathbf{f} = (f_1, \ldots, f_n)^T$ where $f_i \in L^2(E, \rho), i = 1, \ldots, n$ then

$$T\mathbf{f} := (Tf_1, \ldots, Tf_n)^T.$$

As a consequence of this notation we observe that for any $n \times n$ matrix A with scalar entries we have

$$T A\mathbf{f} = AT\mathbf{f}, \quad \mathbf{f} = (f_1, \ldots, f_n)^T, \quad f_i \in L^2(E, \rho), \quad i = 1, 2, \ldots, n.$$

That is, T and A *commute*.

It is difficult in general to construct refinable vector fields for a given set of maps $\phi_0, \ldots, \phi_{m-1}$. Nonetheless, there are important examples in which this can be done. For instance, whenever $\psi_1, \ldots, \psi_{m-1}$ are *affine* maps and $\{p_1, \ldots, p_n\}$ forms a basis for polynomials on $\mathbb{R}^d$ of some total degree N (so that $n = \binom{N+d}{d}$) then the corresponding vector field $\mathbf{p} = (p_1, \ldots, p_n)^T$ is refinable relative to the maps $\psi_0, \ldots, \psi_{m-1}$. Also, when $\phi_i := \phi^{-1} \circ \psi_i \circ \phi, i = 0, 1, \ldots, m-1$ where $\psi_i, i = 0, 1, \ldots, m-1$ are affine mappings but ϕ is otherwise an arbitrary mapping then $\mathbf{f} := \mathbf{p} \circ \phi$ is likewise a refinable vector field relative to the mappings $\phi_i, i = 0, 1, \ldots, m-1$.

The question of the existence of a refinable vector field for a given set of mappings $\phi_0, \ldots, \phi_{m-1}$ and corresponding matrices $A_0, \ldots, A_{m-1}$ will be studied elsewhere. Below we present a sufficient condition which insures the existence of a vector field $\mathbf{f} \in L^\infty(E)$ which satisfies the refinement equations

$$\mathbf{f} \circ \phi_i = A_i^T \mathbf{f}, \quad i = 0, 1, \ldots, m-1, \tag{3.3}$$

and the condition

$$\mathbf{v}^T \mathbf{f} = 1, \, v = (1, \ldots, 1)^T. \tag{3.4}$$

To prepare for this result we make a preliminary remark. We follow [16] and consider the $(n-1) \times n$ *difference matrix* D defined by

$$(D\mathbf{x})_i = x_{i+1} - x_i, \quad i = 1, \ldots, n-1, \quad \mathbf{x} = (x_1, \ldots, x_n)^T.$$

Whenever A is an $n \times n$ matrix such that

$$A\mathbf{v} = \mathbf{v}, \quad \mathbf{v} = (1, \ldots, 1)^T$$

then there is an $(n-1) \times (n-1)$ matrix $\mathcal{A}$ such that

$$\mathcal{A}D = DA. \tag{3.5}$$

We use $\|A\|_2$ to denote the matrix norm

$$\|A\|_2 := \max\{\|A\mathbf{x}\|_2 : \|\mathbf{x}\|_2 \leq 1, \mathbf{x} \in \mathbb{R}^n\},$$

where

$$\|\mathbf{x}\|_2^2 = \sum_{i=1}^{n} x_i^2, \quad \mathbf{x} := (x_1, \ldots, x_n)^T \in \mathbb{R}^n.$$

THEOREM 3.1 *Suppose the mappings $\phi_0, \phi_1, \ldots, \phi_{m-1}$ described in Section 2 have the property that there exists a vector field $\mathbf{h} : E \to \mathbb{R}^n$ in $L^\infty(E)$ and $n \times n$ matrices B_i, $i = 0, 1, \ldots, m-1$ such that*

$$\mathbf{h} \circ \phi_i = B_i^T \mathbf{h}, \quad i = 0, 1, \ldots, m-1, \tag{3.6}$$

$$\mathbf{v}^T \mathbf{h} = 1, \tag{3.7}$$

and

$$B_i \mathbf{v} = \mathbf{v}, \quad i = 0, 1, \ldots, m-1. \tag{3.8}$$

Let $A_0, A_1, \ldots, A_{m-1}$ be $n \times n$ matrices such that

$$A_i \mathbf{v} = \mathbf{v}, \quad i = 0, 1, \ldots, m-1 \tag{3.9}$$

and there exist constants $M > 0$, $\lambda \in (0, 1)$ such that for all $k \geq 1$, $i_1, \ldots, i_k \in \{0, 1, \ldots, m-1\}$

$$\|\mathcal{A}_{i_k} \cdots \mathcal{A}_{i_1}\|_2 \leq M\lambda^k. \tag{3.10}$$

Then there is a vector field $\mathbf{f} : E \to \mathbb{R}^n$ in $L^\infty(E)$ which satisfies equations (3.3) and (3.4).

Proof: We follow the proof of Theorem 1.2 from [16] and introduce a sequence of vector fields $\mathbf{f}_k \in L^\infty(E), k = 0, 1, \ldots$ by setting

$$\mathbf{f}_0 := \mathbf{h} \tag{3.11}$$

and

$$\mathbf{f}_k \circ \phi_i = A_i^T \mathbf{f}_{k-1}, \quad i = 0, 1, \ldots, m - 1. \tag{3.12}$$

From these equations it follows that

$$\mathbf{f}_k((\phi_{i_1} \circ \cdots \circ \phi_{i_k})(x)) = A_{\mathbf{e}^k}^T \mathbf{h}(x), \quad x \in E, \tag{3.13}$$

for any $i_1, \ldots, i_k \in \{0, 1, \ldots, m - 1\}$ where we set

$$A_{\mathbf{e}^k} := A_{i_k} \cdots A_{i_1}. \tag{3.14}$$

Choose any $\sigma \in E$. It will be proved later in Theorem 4.7 (see equation (4.19)) that for every $k \geq 0$ there is a $t \in E$ and $i_1, \ldots, i_{k+1} \in \{0, 1, \ldots, m - 1\}$ such that

$$\sigma = (\phi_{i_1} \circ \cdots \circ \phi_{i_{k+1}})(t). \tag{3.15}$$

Let $s := \phi_{i_{k+1}}(t)$ and so it follows that

$$\sigma = (\phi_{i_1} \circ \cdots \circ \phi_{i_k})(s). \tag{3.16}$$

Therefore, from equations (3.13) and (3.6) it follows that

$$\begin{aligned}
\mathbf{f}_{k+1}(\sigma) - \mathbf{f}_k(\sigma) &= A_{\mathbf{e}^{k+1}}^T \mathbf{h}(t) - A_{\mathbf{e}^k}^T \mathbf{h}(s) \\
&= A_{\mathbf{e}^k}^T A_{i_{k+1}}^T \mathbf{h}(t) - A_{\mathbf{e}^k}^T B_{i_{k+1}}^T \mathbf{h}(t) \\
&= (S_{i_{k+1}} A_{\mathbf{e}^k})^T \mathbf{h}(t)
\end{aligned}$$

where we introduce the $n \times n$ matrices

$$S_i := A_i - B_i, \quad i = 0, 1, \ldots, m - 1.$$

Since

$$S_i \mathbf{v} = 0, \quad i = 0, 1, \ldots, m - 1$$

there are $n \times (n - 1)$ matrices $R_i, i = 0, 1, \ldots, m - 1$ such that

$$S_i = R_i D, \quad i = 0, 1, \ldots, m - 1. \tag{3.17}$$

Hence by equation (3.5) we have

$$\mathbf{f}_{k+1}(\sigma) - \mathbf{f}_k(\sigma) = (R_{i_{k+1}} A_{i_k} \cdots A_{i_1} D)^T \mathbf{h}(t) \tag{3.18}$$

and so by (3.10) there is a constant $K > 0$ such that

$$\|\mathbf{f}_{k+1}(\sigma) - \mathbf{f}_k(\sigma)\|_2 \leq K\lambda^k \|\mathbf{h}(t)\|_2, \quad k = 1, 2, \ldots, \quad \sigma \in E.$$

Since $\mathbf{h} \in L^\infty(E)$ it follows that $\{\mathbf{f}_k : k = 0, 1, \ldots\}$ is a Cauchy sequence in $L^\infty(E)$. Let $\mathbf{f}$ be its limit. Then equation (3.12) implies that

$$\mathbf{f} \circ \phi_i = A_i^T \mathbf{f}, \quad i = 0, 1, \ldots, m - 1$$

while equations (3.7) and (3.13) yields the equations

$$\mathbf{v}^T \mathbf{f}_k = 1, \quad k = 1, 2 \ldots.$$

Hence in the limit $\mathbf{v}^T \mathbf{f} = 1$. ∎

COROLLARY 3.2 *Suppose that $\phi_0, \ldots, \phi_{m-1}$ are the mappings described in Section 2. Let $A_0, A_1, \ldots, A_{m-1}$ be $n \times n$ matrices satisfy equation (3.9) and there exist constants $M > 0$, $\lambda \in (0, 1)$ such that for all $k \geq 1$, $i_1, \ldots, i_k \in \{0, 1, \ldots, m - 1\}$ (3.10) holds. Then there is a vector field $\mathbf{f} : E \to \mathbb{R}^n$ in $L^\infty(E)$ which satisfies equations (3.3) and (3.4).*

Proof: In Theorem 3.1 choose

$$\mathbf{h} = \left(\frac{1}{n}, \frac{1}{n}, \ldots, \frac{1}{n}\right)^T$$

and

$$B_i = I, \quad i = 0, 1, \ldots, m - 1.$$

Then equations (3.6) to (3.8) hold. Therefore, this corollary follows directly from Theorem 3.1. ∎

From this corollary we see that the existence of a vector field which satisfies equations (3.3) and (3.4) is independent of the mappings $\phi_0, \phi_1, \ldots, \phi_{m-1}$. It only depends on the refinement matrices.

It is important to keep in mind that the point of view we describe in this paper is based first on choosing the mappings ϕ_i, $i = 0, 1, \ldots, m - 1$ so that the operators P_i, $i = 0, 1, \ldots, m - 1$, in (2.3) are isometries on $L^2(E, \rho)$ and then on choosing a refinable vector field $\mathbf{f}$ relative to ϕ_i, $i = 0, 1, \ldots, m - 1$. However, later when we adopt a strictly matrix point of view and develop reconstruction and decomposition algorithms we will see that it is *only* the refinement matrices $A_0, \ldots, A_{m-1}$ which are needed. That is, these reconstruction and decomposition algorithms depend neither on the mappings ϕ_i, $i = 0, 1, \ldots, m - 1$ nor the refinable vector field $\mathbf{f}$ but *only* on the matrices $A_0, \ldots, A_{m-1}$.

4. Recursive Generation of Biorthogonal Bases in $L^2(E, \rho)$

In this section, we describe a recursive procedure for generating biorthogonal bases in $L^2(E, \rho)$. To begin our recursion, we start with two refinable vector fields $\mathbf{f} = (f_1, \ldots, f_n)^T$ and $\tilde{\mathbf{f}} = (\tilde{f}_1, \ldots, \tilde{f}_n)^T$, with $f_i, \tilde{f}_i \in L^2(E, \rho)$, $i = 1, \ldots, n$, that satisfy the matrix refinement equation

$$G_i \mathbf{f} = \sqrt{\gamma_i} A_i^T \mathbf{f}, \quad i = 0, 1, \ldots, m - 1$$

and

$$G_i \tilde{\mathbf{f}} = \sqrt{\gamma_i} \tilde{A}_i^T \tilde{\mathbf{f}}, \quad i = 0, 1, \ldots, m - 1.$$

Here $A_i, \tilde{A}_i, i = 0, 1, \ldots, m - 1$ are some prescribed $n \times n$ matrices and

$$\mathbf{v}^T \mathbf{f} = a, \quad \tilde{\mathbf{v}}^T \tilde{\mathbf{f}} = \tilde{a},$$

where $\mathbf{v}, \tilde{\mathbf{v}}$ are some constant vectors in $\in \mathbb{R}^n$ and $a, \tilde{a}$ are some *nonzero* constants. We assume that the components of $\mathbf{f}$ and $\tilde{\mathbf{f}}$ are chosen so that they satisfy the additional condition

$$\langle f_i, \tilde{f}_j \rangle = \delta_{ij}, \quad i, j = 1, 2 \ldots, n. \tag{4.1}$$

For us the pair of sets $\mathbb{F}_0 = \{f_1, \ldots, f_n\}$ and $\tilde{\mathbb{F}}_0 = \{\tilde{f}_1, \ldots, \tilde{f}_n\}$ form our *initial biorthogonal pair* in $L^2(E, \rho)$.

We shall say two finite sets $\mathbb{F}_0, \tilde{\mathbb{F}}_0$ are a *biorthogonal pair* provided that they satisfy condition (4.1). Note that this condition is *dependent* on the way we order the elements in $\mathbb{F}_0, \tilde{\mathbb{F}}_0$. From these sets we will generate a *countable* scale of biorthogonal subspaces of $L^2(E, \rho)$.

To this end, we review some standard notation which will be convenient for us to use. If R and S are subspaces of $L^2(E, \rho)$ we say that R is *orthogonal* to S, denoted by $R \perp S$, whenever $\langle f, g \rangle = 0$ for all $f \in R$ and $g \in S$. When $R \cap S = \{0\}$ we use $R \oplus S$ to denote the subspace of $L^2(E, \rho)$ formed by the *direct sum* of R and S given by

$$R \oplus S := \{f + g : \ f \in R, \ g \in S\}.$$

In the case that R is orthogonal to S, we use the notation $R \oplus^\perp S$ to indicate that this is an *orthogonal direct sum*. Observe that whenever T is any one to one bounded linear operator on $L^2(E, \rho)$ then

$$T(R \oplus S) = TR \oplus TS.$$

Moreover, if T is an isometry then $TR \perp TS$ if and ony if $R \perp S$. Therefore, we conclude in the case that

$$T(R \oplus^\perp S) = TR \oplus^\perp TS.$$

Obviously the operations $\oplus, \oplus^\perp$ are commutative and associative. Furthermore, we observe that the linear operators $T_i, i = 0, 1, \ldots, m - 1$ defined by (2.13) have the property that for any subset R of $L^2(E, \rho)$, it follows that

$$T_i(R) \perp T_j(R), \quad i \neq j, \quad i, j = 0, 1, \ldots, m - 1.$$

Let us use these facts to generate *recursively* a scale of subspace in $L^2(E, \rho)$ by using the operation of orthogonal direct sum of subspaces. First, we set

$$F_0 := \{\mathbf{c}^T \mathbf{f} : \mathbf{c} \in \mathbb{R}^n\}$$

and then

$$F_{k+1} := \bigoplus_{i=0}^{m-1} {}^{\perp} T_i F_k, \quad k = 0, 1, \ldots. \tag{4.2}$$

Similarly, we define

$$\tilde{F}_0 = \{\mathbf{c}^T \tilde{\mathbf{f}} : \mathbf{c} \in \mathbb{R}^n\}$$

and

$$\tilde{F}_{k+1} = \bigoplus_{i=0}^{m-1} {}^{\perp} T_i \tilde{F}_k, \quad k = 0, 1, \ldots. \tag{4.3}$$

Our goal is to use these pair of scales of subspaces in $L^2(E, \rho)$ to generate, recursively in k, a scale of biorthogonal subspaces in $L^2(E, \rho)$. Before doing this we need to record some properties of F_k and $\tilde{F}_k$. In particular, the next lemma insures that these subspaces are nested.

LEMMA 4.1 *Let F_k and $\tilde{F}_k$ be defined as above. Then*

$$F_k \subseteq F_{k+1}, \quad \tilde{F}_k \subseteq \tilde{F}_{k+1}, \quad k = 0, 1, \ldots, \tag{4.4}$$

and

$$T_i^* F_{k+1} \subseteq F_k, \quad T_i^* \tilde{F}_{k+1} \subseteq \tilde{F}_k, \quad i = 0, 1, \ldots, m - 1, \quad k = 0, 1, \ldots. \tag{4.5}$$

Proof: We only present the proof for F_k since the statements for $\tilde{F}_k$ can be proved similarly.

We proceed to prove (4.4) by induction on k. By the definition of the linear operator G_i and the refinability of the vector field $\mathbf{f}$ we have

$$G_i \mathbf{f} = \sqrt{\gamma_i} A_i^T \mathbf{f}, \quad i = 0, 1, \ldots, m - 1. \tag{4.6}$$

Applying H_i to both sides of (4.6) and summing the resulting equations over $i = 0, 1, \ldots,$ $m - 1$, we conclude from Lemma 2.2 that

$$\sum_{i=0}^{m-1} \sqrt{\gamma_i} H_i A_i^T \mathbf{f} = \sum_{i=0}^{m-1} H_i G_i \mathbf{f} = \mathbf{f}.$$

Since H_i and A_i commute, we have established that

$$\sum_{i=0}^{m-1} \sqrt{\gamma_i} A_i^T H_i \mathbf{f} = \mathbf{f}.$$

Equivalently, we have

$$\sum_{j=0}^{m-1} \left(\sum_{i=0}^{m-1} \sqrt{\gamma_i} q_{ji} A_i^T \right) T_j \mathbf{f} = \mathbf{f},$$

from which it follows that $F_0 \subseteq F_1$.

Our induction hypothesis leads us to suppose that $F_\ell \subseteq F_{\ell+1}$, $\ell \geq 0$ and for notational convenience we set $k := \ell + 1$. Since

$$F_k = \bigoplus_{i=0}^{m-1} {}^\perp T_i F_\ell \subseteq \bigoplus_{i=0}^{m-1} {}^\perp T_i F_{\ell+1} = F_{k+1}$$

we have advanced the induction hypothesis thereby establishing that F_k, $k = 0, 1, \ldots,$ are indeed nested in k.

To establish the inclusion (4.5), in view of equation (2.14) it suffices to verify that

$$G_j F_{k+1} \subseteq F_k, \quad j = 0, 1, \ldots, m - 1, \quad k = 0, 1, \ldots.$$

To prove this fact, we choose any $g \in F_{k+1}$. Then by the definition of F_{k+1} there are m functions $g_i \in F_k$, $i = 0, 1, \ldots, m - 1$ such that

$$g = \sum_{i=0}^{m-1} T_i g_i.$$

Applying the linear operators G_j, $j = 0, 1, \ldots, m - 1$ to both sides of the above equation we obtain

$$G_j g = \sum_{i=0}^{m-1} G_j T_i g_i.$$

Using equation (2.13) and then (2.10)–(2.11) we conclude that

$$G_j g = \sum_{i=0}^{m-1} G_j \sum_{\ell=0}^{m-1} q_{i\ell} P_\ell g_i$$

$$= \sum_{i=0}^{m-1} \sum_{\ell=0}^{m-1} q_{i\ell} P_j^* P_\ell g_i$$

$$= \sum_{i=0}^{m-1} q_{ij} g_i.$$

Thus, in particular, we conclude that $G_j g \in F_k$. ∎

The following elementary fact will also be useful.

LEMMA 4.2 *Suppose that the sets* $\mathbb{G} := \{g_1, \ldots, g_N\}$ *and* $\mathbb{H} := \{h_1, \ldots, h_N\}$ *are a biorthogonal pair. Then the sets*

$$\mathcal{G} := \{T_0 g_1, \ldots, T_0 g_N, T_1 g_1, \ldots, T_1 g_N, \ldots, T_{m-1} g_1, \ldots, T_{m-1} g_N\}$$

and

$$\mathcal{H} := \{T_0 h_1, \ldots, T_0 h_N, T_1 h_1, \ldots, T_1 h_N, \ldots, T_{m-1} h_1, \ldots, T_{m-1} h_N\}$$

are likewise a biogthogonal pair.

The next lemma is tailor made for our purposes although it is actually valid in greater generality.

LEMMA 4.3 *Suppose A, B, C are finite dimensional subspaces of some Hilbert space such that* $B \subseteq A$ *and B, C have bases* $\mathbb{B} = \{b_1, \ldots, b_N\}$, $\mathbb{C} = \{c_1, \ldots, c_N\}$ *respectively, such that* $\mathbb{B}$ *and* $\mathbb{C}$ *are a biorthogonal pair. Then there is a unique subspace* $D \subseteq A$ *such that* $D \perp C$ *and* $B \oplus D = A$.

Remark. The subspace D may be thought of as the orthogonal complement of B in A relative to C.

We are now ready to introduce our scale of biorthogonal subspaces in $L^2(E, \rho)$. According to the definition of the subspaces F_k and $\tilde{F}_k$ we see by applying Lemma 4.2 inductively in k, that when $\mathbb{F}_0$ and $\tilde{\mathbb{F}}_0$ are a biorthogonal pair, then F_k and $\tilde{F}_k$ have bases which likewise are a biorthogonal pair. Hence we may apply Lemma 4.3 to the subspaces $A = F_{k+1}$, $B = F_k$ and $C = \tilde{F}_k$ to conclude that there is a unique subspace W_k of F_{k+1} such that

$$F_k \oplus W_k = F_{k+1}, \quad k = 0, 1, \ldots, \tag{4.7}$$

$$W_k \perp \tilde{F}_k, \quad k = 0, 1, \ldots. \tag{4.8}$$

Similarly, for $A = \tilde{F}_{k+1}$, $B = \tilde{F}_k$ and now $C = F_k$ we obtain a unique subspace $\tilde{W}_k$ of $\tilde{F}_{k+1}$ such that

$$\tilde{F}_k \oplus \tilde{W}_k = \tilde{F}_{k+1}, \quad k = 0, 1, \ldots,$$

and

$$\tilde{W}_k \perp F_k, \quad k = 0, 1, \ldots.$$

Since both F_0 and $\tilde{F}_0$ are of dimension n, we have

$$\dim F_k = \dim \tilde{F}_k = m^k n. \tag{4.9}$$

It follows from (4.9) and the definition of W_k, $\tilde{W}_k$ that

$$\dim W_k = \dim \tilde{W}_k = m^k (m - 1)n. \tag{4.10}$$

In the next theorem we demonstrate that the spaces W_k and $\tilde{W}_k$ can be easily generated *recursively* in k. To this end, we let

$$W_0 := \operatorname{span}\mathbb{W}_0, \quad \tilde{W}_0 := \operatorname{span}\tilde{\mathbb{W}}_0,$$

where

$$\mathbb{W}_0 := \{w_1, \ldots, w_{(m-1)n}\}$$

and

$$\tilde{\mathbb{W}}_0 := \{\tilde{w}_1, \ldots, \tilde{w}_{(m-1)n}\}$$

are chosen so that the functions w_i, $\tilde{w}_i$, $i = 1, \ldots, (m - 1)n$ form a basis for W_0, $\tilde{W}_0$, respectively.

THEOREM 4.4 *Assume that the sets* $\mathbb{F}_0$, $\tilde{\mathbb{F}}_0$ *and* $\mathbb{W}_0$, $\tilde{\mathbb{W}}_0$ *are both biorthogonal pairs such that*

$$W_0 \perp \tilde{F}_0, \quad \tilde{W}_0 \perp F_0. \tag{4.11}$$

Then

$$W_{k+1} = \bigoplus_{i=0}^{m-1} {}^\perp T_i W_k, \quad k = 0, 1, \ldots, \tag{4.12}$$

and

$$\tilde{W}_{k+1} = \bigoplus_{i=0}^{m-1} {}^\perp T_i \tilde{W}_k, \quad k = 0, 1, \ldots. \tag{4.13}$$

Proof: We prove this theorem by induction on k. It is only necessary to prove (4.12) as the same argument proves (4.13). Let us first prove equation (4.12) when $k = 0$. That is, we wish to verify that

$$F_2 = F_1 \oplus \left(\bigoplus_{i=0}^{m-1} {}^{\perp} T_i W_0 \right),$$

(4.14)

and

$$\left(\bigoplus_{i=0}^{m-1} {}^{\perp} T_i W_0 \right) \perp \tilde{F}_1.$$

(4.15)

By the definition of F_2 and our assumption, we have

$$\begin{aligned}
F_2 &= \bigoplus_{i=0}^{m-1} {}^{\perp} T_i F_1 \\
&= \left(\bigoplus_{i=0}^{m-1} {}^{\perp} T_i F_0 \right) \oplus \left(\bigoplus_{i=0}^{m-1} {}^{\perp} T_i W_0 \right) \\
&= F_1 \oplus \left(\bigoplus_{i=0}^{m-1} {}^{\perp} T_i W_0 \right).
\end{aligned}$$

Hence, equation (4.14) is proved. Since $W_0 \perp \tilde{F}_0$, we have $T_i W_0 \perp T_j \tilde{F}_0$ for $i, j = 0, 1, \ldots,$ $m - 1$. Therefore, relation (4.15) follows.

To complete the proof, we assume the statements of the theorem hold for $k = \ell$. That is, we have

$$F_{\ell+2} = F_{\ell+1} \oplus \left(\bigoplus_{i=0}^{m-1} {}^{\perp} T_i W_\ell \right)$$

(4.16)

and

$$W_{\ell+1} = \bigoplus_{i=0}^{m-1} {}^{\perp} T_i W_\ell \perp \tilde{F}_{\ell+1}.$$

(4.17)

We wish to prove equations (4.16) and (4.17) even when ℓ is replaced by $\ell + 1$. By the definition of $F_{\ell+3}$ and equation (4.17), we have

$$\begin{aligned}
F_{\ell+3} &= \bigoplus_{i=0}^{m-1} {}^{\perp} T_i F_{\ell+2} \\
&= \left(\bigoplus_{i=0}^{m-1} {}^{\perp} T_i F_{\ell+1} \right) \oplus \left(\bigoplus_{i=0}^{m-1} {}^{\perp} T_i W_{\ell+1} \right) \\
&= F_{\ell+2} \oplus \left(\bigoplus_{i=0}^{m-1} {}^{\perp} T_i W_{\ell+1} \right).
\end{aligned}$$

Hence, we have proved equation (4.16) with ℓ replaced by $\ell + 1$. To prove relation (4.17) holds when ℓ is replaced by $\ell + 1$, we observe that (4.17) implies that

$$T_i W_{\ell+1} \perp T_j \tilde{F}_{\ell+1}, \quad i, j = 0, 1, \ldots, m - 1.$$

From this observation equation (4.17) holds when ℓ is replaced by $\ell + 1$. ∎

THEOREM 4.5 *Under the hypotheses of Theorem 4.4, the following statements hold:*
  (i) $F_0 \perp \tilde{W}_k$, $k = 0, 1, \ldots$.
  (ii) $\tilde{F}_0 \perp W_k$, $k = 0, 1, \ldots$.
  (iii) $W_k \perp \tilde{W}_{k'}$, $k \neq k'$.

Proof: By the definition of $\tilde{W}_0$ we have that $F_0 \perp \tilde{W}_0$. When $k \geq 1$ we have $F_0 \subset F_k$ and $F_k \perp \tilde{W}_k$, and so it follows that $F_0 \perp \tilde{W}_k$. In an identical manner we prove that (ii) holds.

As for (iii) we let k' be a fixed non-negative integer. For $k < k'$, we have

$$\tilde{W}_{k'} \perp F_{k'} \supseteq F_{k+1} = F_k \oplus W_k \supseteq W_k,$$

it follows that $\tilde{W}_{k'} \perp W_k$ in this case. For $k > k'$, we observe that

$$\tilde{W}_{k'} \subseteq \tilde{F}_{k'} \oplus \tilde{W}_{k'} = \tilde{F}_{k'+1} \subseteq \tilde{F}_k \perp W_k,$$

to conclude that $\tilde{W}_{k'} \perp W_k$ in this case as well. ∎

The next theorem demonstrates how easily an orthonormal basis can be generated for both W_k and $\tilde{W}_k$. To this end we denote

$$w_{-1,i} = f_i, \quad i = 1, 2, \ldots, n,$$

$$w_{0,i} = w_i, \quad i = 1, 2, \ldots, (m-1)n,$$

and

$$w_{j+1, i+\ell(m-1)m^j n} = T_\ell w_{j,i}, \quad i = 1, 2, \ldots, m^j (m-1)n, \quad \ell = 0, 1, \ldots, m-1, \quad j = 0, 1, \ldots.$$

Similarly, we define $\tilde{w}_{-1,i}$, $i = 1, 2, \ldots, n$ and $\tilde{w}_{j,i}$ for $i = 1, 2, \ldots, m^j (m-1)n$, $j = 0, 1, \ldots$.

THEOREM 4.6 *Under the hypotheses of Theorem 4.4, there holds the formula*

$$\langle w_{j,i}, \tilde{w}_{j',i'} \rangle = \delta_{j,j'} \delta_{i,i'},$$

$i = 1, 2, \ldots, m^j (m-1)n$, $j = 0, 1, \ldots$, $i' = 1, 2, \ldots, m^{j'} (m-1)n$, $j' = 0, 1, \ldots$.

$$(4.18)$$

Proof: This result follows directly from Lemma 4.2. ■

Our next result demonstrates under some additional hypotheses that each of the scale of spaces $\{F_k\}$, $\{\tilde{F}_k\}$, $k = 0, 1, \ldots$, forms a *multiresolution*. To this end, we introduce the following notation. Let k be a positive integer. For every $\epsilon_1, \ldots, \epsilon_k \in \{0, 1, \ldots, m - 1\}$ we set

$$\mathbf{e} := (\epsilon_1, \ldots, \epsilon_k)^T \in \{0, 1, \ldots, m - 1\}^k.$$

We define the set

$$E_\mathbf{e} = \phi_\mathbf{e}(E) := (\phi_{\epsilon_1} \circ \cdots \circ \phi_{\epsilon_k})(E)$$

and the diameter

$$\delta_k(E) := \max\{\text{diam } E_\mathbf{e} : \ \mathbf{e} \in \{0, 1, \ldots, m - 1\}^k\}$$

where for any set $S \subseteq \mathbb{R}^d$

$$\text{diam } S := \sup\{\|x - y\|_2 : x, y \in S\}.$$

THEOREM 4.7 *Suppose that the mappings* ϕ_i, $i = 0, 1, \ldots, m - 1$ *are Hölder continuous and*

$$\lim_{k \to \infty} \delta_k(E) = 0.$$

Then the following statements hold:

(i) $cl_{L^2(E,\rho)} \left(\bigcup_{i=0}^\infty F_i \right) = cl_{L^2(E,\rho)} \left(\bigcup_{i=0}^\infty \tilde{F}_i \right) = L^2(E, \rho).$

(ii) $cl_{L^2(E,\rho)} \left(F_0 \oplus \bigoplus_{i=0}^\infty W_i \right) = cl_{L^2(E,\rho)} \left(\tilde{F}_0 \oplus \bigoplus_{i=0}^\infty \tilde{W}_i \right) = L^2(E, \rho).$

Proof: Since (ii) is a consequence of (i) it suffices to prove this claim. To this end, we first point out that for each positive integer k

$$E = \bigcup \{E_\mathbf{e} : \mathbf{e} \in \{0, 1, \ldots, m - 1\}^k\} \tag{4.19}$$

and

$$\text{meas} \, (E_\mathbf{e} \cap E_{\mathbf{e}'}) = 0, \ \ \mathbf{e} \neq \mathbf{e}', \ \mathbf{e}, \mathbf{e}' \in \{0, 1, \ldots, m - 1\}^k. \tag{4.20}$$

Each of these facts follow by induction on k. For instance, to prove (4.19) we use the formula that

$$E = \bigcup_{i=0}^{m-1} \phi_i(E)$$

to conclude that

$$\phi_{\mathbf{e}}(E) = \bigcup_{i=0}^{m-1} \phi_{\mathbf{e}_i}(E), \quad \mathbf{e} \in \{0, 1, \ldots, m-1\}^k$$

where $\mathbf{e}_i := (\mathbf{e}, i) \in \{0, 1, \ldots, m-1\}^{k+1}$, $i = 0, 1, \ldots, m-1$. From this observation (4.19) follows by induction on k.

We shall also prove (4.20) by induction on k. To this end, suppose (4.20) is valid. We will advance the induction step by considering $\mathbf{e} = (\epsilon_1, \mathbf{e}^1)$, $\tilde{\mathbf{e}} = (\tilde{\epsilon}_1, \tilde{\mathbf{e}}^1)$ with $\mathbf{e} \neq \tilde{\mathbf{e}}$ where $\mathbf{e}, \tilde{\mathbf{e}} \in \{0, 1, \ldots, m-1\}^{k+1}$, $\epsilon_1, \tilde{\epsilon}_1 \in \{0, 1, \ldots, m-1\}$ and $\mathbf{e}^1, \tilde{\mathbf{e}}^1 \in \{0, 1, \ldots, m-1\}^k$. Clearly, it is the case that

$$E_{\mathbf{e}} \cap E_{\tilde{\mathbf{e}}} = \phi_{\epsilon_1}(R) \cap \phi_{\tilde{\epsilon}_1}(S)$$

where $R := E_{\mathbf{e}^1}$ and $S := E_{\tilde{\mathbf{e}}^1}$. When $\epsilon_1 = \tilde{\epsilon}_1$ we conclude that

$$E_{\mathbf{e}} \cap E_{\tilde{\mathbf{e}}} = \phi_{\epsilon_1}(R \cap S)$$

and so because by hypothesis ϕ_{ϵ_1} is Hölder continuous and meas $(R \cap S) = 0$ it follows that meas $(E_{\mathbf{e}} \cap E_{\tilde{\mathbf{e}}}) = 0$. When $\epsilon_1 \neq \tilde{\epsilon}_1$ then

$$E_{\mathbf{e}} \cap E_{\tilde{\mathbf{e}}} \subseteq E_{\epsilon_1} \cap E_{\tilde{\epsilon}_1}$$

and so again meas $(E_{\mathbf{e}} \cap E_{\tilde{\mathbf{e}}}) = 0$ because by definition meas $(E_{\epsilon_1} \cap E_{\tilde{\epsilon}_1}) = 0$.

Now that we have established (4.19) and (4.20) we may follow the reasoning used in the proof of Theorem 4.2 of [1] to assert that

$$F_k = \text{span}\,\{g : g|_{E_{\mathbf{e}}} \in F_k, \ \mathbf{e} \in \{0, 1, \ldots, m-1\}^k\}$$

where $g|_{E_{\mathbf{e}}}$ denotes the function obtained by restricting g to the set $E_{\mathbf{e}}$. In particular, F_k contains all functions which are piecewise constant on the sets $E_{\mathbf{e}}$, $\mathbf{e} \in \{0, 1, \ldots, m-1\}^k$ because of property (3.2) of refinable vector fields. In view of the hypothesis that $\lim_{k \to \infty} \delta_k(E) = 0$ the result follows. ∎

5. Constructions of the Initial Wavelet Space

According to Theorem 4.6 to generate our biorthogonal bases in $L^2(E, \rho)$ we need only identify such bases for W_0 and $\tilde{W}_0$. In this section, we study the construction of bases for the initial biorthogonal wavelet spaces W_0 and $\tilde{W}_0$. Let A_i and $\tilde{A}_i$, $i = 0, 1, \ldots, m-1$ be $n \times n$ matrices which have associated nontrivial refinable vector fields $\mathbf{f} = (f_1, \ldots, f_n)^T$ and $\tilde{\mathbf{f}} = (\tilde{f}_1, \ldots, \tilde{f}_n)^T$, respectively, such that the sets $\mathbb{F}_0$ and $\tilde{\mathbb{F}}_0$ are a biorthogonal pair. That is,

$$\mathbf{f}(\phi_i(t)) = A_i^T \mathbf{f}(t), \quad t \in E \tag{5.1}$$

$$\tilde{\mathbf{f}}(\phi_i(t)) = \tilde{A}_i^T \tilde{\mathbf{f}}(t), \quad t \in E \tag{5.2}$$

and

$$\int_E \mathbf{f}(t)\tilde{\mathbf{f}}^T(t)d\rho(t) = I_n \tag{5.3}$$

where I_n denotes the $n \times n$ identity matrix.

As in the previous section, we let

$$F_1 = F_0 \oplus W_0, \quad \tilde{F}_1 = \tilde{F}_0 \oplus \tilde{W}_0,$$

where

$$F_0 = \{\mathbf{c}^T \mathbf{f} : \mathbf{c} \in \mathbb{R}^n\}, \quad \tilde{F}_0 = \{\mathbf{c}^T \tilde{\mathbf{f}} : \mathbf{c} \in \mathbb{R}^n\},$$

and

$$W_0 = \{\mathbf{c}^T \mathbf{w} : \mathbf{c} \in \mathbb{R}^{(m-1)n}\}, \quad \tilde{W}_0 = \{\mathbf{c}^T \tilde{\mathbf{w}} : \mathbf{c} \in \mathbb{R}^{(m-1)n}\}.$$

The biorthogonal wavelet vector fields $\mathbf{w}, \tilde{\mathbf{w}} : E \to \mathbb{R}^{(m-1)n}$ necessarily have the following properties:

(a) $\mathbf{w}(\phi_i(t)) = B_i^T \mathbf{f}(t)$ and $\tilde{\mathbf{w}}(\phi_i(t)) = \tilde{B}_i^T \tilde{\mathbf{f}}(t)$, $t \in E$, for some $n \times (m-1)n$ matrices B_i and $\tilde{B}_i$, $i = 0, 1, \ldots, m-1$.

(b) $\int_E \mathbf{w}(t)\tilde{\mathbf{w}}^T(t)d\rho(t) = I_{(m-1)n}$.

(c) $\int_E \mathbf{w}(t)\tilde{\mathbf{f}}^T(t)d\rho(t) = O_{(m-1)n \times n}$ and $\int_E \tilde{\mathbf{w}}(t)\mathbf{f}^T(t)d\rho(t) = O_{(m-1)n \times n}$.

Keep in mind that each A_i is a square matrix of order n while B_i is a rectangular matrix of size $n \times (m-1)n$, $i = 0, 1, \ldots, m-1$.

LEMMA 5.1 *Let* $\mathbf{w}, \tilde{\mathbf{w}} : E \to \mathbb{R}^{(m-1)n}$ *be biorthogonal wavelet vector fields. Then they satisfy conditions (a)–(c) if and only if the matrices B_i and $\tilde{B}_i$ satisfy the matrix equations*

$$\sum_{i=0}^{m-1} \gamma_i B_i^T \tilde{B}_i = I_{(m-1)n}, \tag{5.4}$$

$$\sum_{i=0}^{m-1} \gamma_i B_i^T \tilde{A}_i = 0_{(m-1)n \times n}, \tag{5.5}$$

and

$$\sum_{i=0}^{m-1} \gamma_i A_i^T \tilde{B}_i = 0_{(m-1)n \times n} \tag{5.6}$$

Moreover, equation (5.3) implies the formula

$$\sum_{i=0}^{m-1} \gamma_i A_i^T \tilde{A}_i = I_n. \tag{5.7}$$

Proof: We only present the proof for equation (5.4). Using the formula given in Lemma 2.1 (ii) and equation (5.3), we have

$$
\begin{aligned}
I_{(m-1)n} &= \int_E \mathbf{w}(t)\tilde{\mathbf{w}}^T(t)d\rho(t) \\
&= \sum_{i=0}^{m-1} \gamma_i \int_E \mathbf{w}(\phi_i(t))\tilde{\mathbf{w}}^T(\phi_i(t))d\rho(t) \\
&= \sum_{i=0}^{m-1} \gamma_i \int_E B_i^T \mathbf{f}(t)\tilde{\mathbf{f}}^T(t)\tilde{B}_i d\rho(t) \\
&= \sum_{i=0}^{m-1} \gamma_i B_i^T \tilde{B}_i.
\end{aligned}
$$

The proof for the remaining equations in this lemma is similar.   ∎

Notice that whenever B_i and $\tilde{B}_i$, $i = 0, 1, \ldots, m - 1$ solve equations (5.4)–(5.6), then $B_i V^T$ and $\tilde{B}_i V^{-1}$, $i = 0, 1, \ldots, m - 1$ also solve equations (5.4)–(5.6) whenever V is an $(m - 1)n \times (m - 1)n$ nonsingular matrix. In fact, we show below that *all* solutions of equations (5.4)–(5.6) have this form.

THEOREM 5.2 *Suppose γ_i are positive constants and A_i, $\tilde{A}_i$ are $n \times n$ matrices, $i = 0, 1, \ldots, m - 1$ which satisfy equation (5.7). Let B_i and $\tilde{B}_i$, $i = 0, 1, \ldots, m - 1$ be a particular solution of equations (5.4)–(5.6). Then,* **any** *solution of equations (5.4)–(5.6) has the form*

$$
B_i V^T, \quad \tilde{B}_i V^{-1} \quad i = 0, 1, \ldots, m - 1,
$$

where V is an $(m - 1)n \times (m - 1)n$ nonsingular matrix.

Proof: Define two $mn \times mn$ matrices

$$
R := \begin{pmatrix} \sqrt{\gamma_0}A_0 & \sqrt{\gamma_0}B_0 \\ \vdots & \vdots \\ \sqrt{\gamma_{m-1}}A_{m-1} & \sqrt{\gamma_{m-1}}B_{m-1} \end{pmatrix} \tag{5.8}
$$

and

$$
\tilde{R} := \begin{pmatrix} \sqrt{\gamma_0}\tilde{A}_0 & \sqrt{\gamma_0}\tilde{B}_0 \\ \vdots & \vdots \\ \sqrt{\gamma_{m-1}}\tilde{A}_{m-1} & \sqrt{\gamma_{m-1}}\tilde{B}_{m-1} \end{pmatrix} \tag{5.9}
$$

and notice that equations (5.4)–(5.6) are equivalent to matrix equation

$$
R^T \tilde{R} = I_{mn}. \tag{5.10}
$$

Consequently, both R^T and $\tilde{R}$ are nonsingular and R^T is the *inverse* of $\tilde{R}$. Hence, equations (5.4)–(5.6) are also equivalent to the equation

$$\tilde{R}R^T = I_{mn},$$

which is the same as saying that

$$\begin{pmatrix} \gamma_0(\tilde{A}_0 A_0^T + \tilde{B}_0 B_0^T) & \cdots & \sqrt{\gamma_0\gamma_{m-1}}(\tilde{A}_0 A_{m-1}^T + \tilde{B}_0 B_{m-1}^T) \\ \vdots & & \vdots \\ \sqrt{\gamma_{m-1}\gamma_0}(\tilde{A}_{m-1} A_0^T + \tilde{B}_{m-1} B_0^T) & \cdots & \gamma_{m-1}(\tilde{A}_{m-1} A_{m-1}^T + \tilde{B}_{m-1} B_{m-1}^T) \end{pmatrix} = I_{mn}.$$

In other words, equations (5.4)–(5.6) are equivalent to the equations

$$\tilde{B}_i B_i^T = \frac{1}{\gamma_i} I_n - \tilde{A}_i A_i^T, \quad i = 0, 1, \dots, m-1, \tag{5.11}$$

and

$$\tilde{B}_i B_j^T = -\tilde{A}_i A_j^T, \quad i \neq j, \ i, j = 0, 1, \dots, m-1. \tag{5.12}$$

Hence, if B_i, $\tilde{B}_i$, $i = 0, 1, \dots, m-1$ and B_i', $\tilde{B}_i'$, $i = 0, 1, \dots, m-1$ are two sets of solutions to equations (5.4)–(5.6) by (5.11) and (5.12), we have

$$\tilde{B}_i B_j^T = \tilde{B}_i' B_j'^T, \quad i, j = 0, 1, \dots, m-1. \tag{5.13}$$

We consider the $(m-1)n \times (m-1)n$ square matrices

$$V := \sum_{i=0}^{m-1} \gamma_i B_i^T \tilde{B}_i' \tag{5.14}$$

and

$$\tilde{V} := \sum_{i=0}^{m-1} \gamma_i B_i'^T \tilde{B}_i. \tag{5.15}$$

Using equations (5.13) and (5.4), we find that

$$\begin{aligned} V\tilde{V} &= \sum_{i=0}^{m-1}\sum_{j=0}^{m-1} \gamma_i\gamma_j B_i^T \tilde{B}_i' B_j'^T \tilde{B}_j \\ &= \sum_{i=0}^{m-1}\sum_{j=0}^{m-1} \gamma_i\gamma_j B_i^T \tilde{B}_i B_j^T \tilde{B}_j \\ &= \left(\sum_{i=0}^{m-1} \gamma_i B_i^T \tilde{B}_i\right)\left(\sum_{j=0}^{m-1} \gamma_j B_j^T \tilde{B}_j\right) \\ &= I_{(m-1)n}. \end{aligned}$$

This imples that $\tilde{V}$ is the inverse of V. Multiplying equation (5.14) on the right by $B_j'^T$ gives

$$V B_j'^T = \sum_{i=0}^{m-1} \gamma_i B_i^T \tilde{B}_i' B_j'^T.$$

Again using equations (5.13) and (5.4), we have

$$V B_j'^T = \left(\sum_{i=0}^{m-1} \gamma_i B_i^T \tilde{B}_i \right) B_j^T = B_j^T, \quad j = 0, 1, \ldots, m-1.$$

Thus,

$$B_j = B_j' V^T, \quad j = 0, 1, \ldots, m-1.$$

On the other hand, multiplying equation (5.15) on the left by $\tilde{B}_j'$ gives

$$\tilde{B}_j' V^{-1} = \tilde{B}_j' \tilde{V} = \sum_{i=0}^{m-1} \gamma_i \tilde{B}_j' B_i'^T \tilde{B}_i, \quad j = 0, 1, \ldots, m-1$$

which implies that

$$\tilde{B}_j' V^{-1} = \sum_{i=0}^{m-1} \gamma_i \tilde{B}_j B_i^T \tilde{B}_i = \tilde{B}_j \sum_{i=0}^{m-1} \gamma_i B_i^T \tilde{B}_i = \tilde{B}_j, \quad j = 0, 1, \ldots, m-1. \qquad \blacksquare$$

In the special case that we wish to construct *orthonormal* wavelets we merely choose $\tilde{\mathbf{f}} = \mathbf{f}$ and $\tilde{A}_i = A_i$, $i = 0, 1, \ldots, m-1$. Therefore, we also set $\tilde{B}_i = B_i$, $i = 0, 1, \ldots, m-1$ and $\tilde{\mathbf{w}} = \mathbf{w}$. Equations (5.4)–(5.6) take the special form

$$\sum_{i=0}^{m-1} \gamma_i B_i^T B_i = I_{(m-1)n} \tag{5.16}$$

$$\sum_{i=0}^{m-1} \gamma_i B_i^T A_i = O_{(m-1)n \times m} \tag{5.17}$$

and (5.7) becomes

$$\sum_{i=0}^{m-1} \gamma_i A_i^T A_i = I_n \tag{5.18}$$

In this case, Theorem 5.2 takes the form

THEOREM 5.3 *Let A_i, $i = 0, 1, \ldots, m-1$ be $n \times n$ matrices and γ_i, $i = 0, 1, \ldots, m-1$ positive constants which satisfy equation (5.18). Let B_i, $i = 0, 1, \ldots, m-1$ be a particular* *solution of (5.16) and (5.17). Then* **any** *solution of (5.16) and (5.17) has the form* $B_i V$, $i = 0, 1, \ldots, m - 1$ *where* V *is an* $(m - 1)n \times (m - 1)n$ *orthogonal matrix.*

Proof: The proof follows by specializing the proof of Theorem 5.2 to the currecnt situation. In particular, we have $\tilde{B}_i = B_i$, $\tilde{B}_i' = B_i'$, $i = 0, 1, \ldots, m - 1$ and so the matrices V, $\tilde{V}$ in equation (5.14), (5.15) are easily seen to have the additional property that $\tilde{V} = V^T$ and so V is in fact an orthogonal matrix. ∎

Returning to the general case, let us next construct a particular solution to the matrix equations (5.4)–(5.6). The first thing to notice is that each of the $mn \times n$ matrices

$$A := (A_0^T, \ldots, A_{m-1}^T)^T$$

and

$$\tilde{A} := (\tilde{A}_0^T, \ldots, \tilde{A}_{m-1}^T)^T$$

are of full rank n in view of equation (5.7). In fact, if $c \in \mathbb{R}^n$ and $A_i c = 0$, $i = 0, 1, \ldots, m-1$ then

$$c = I_n c = \sum_{i=0}^{m-1} \gamma_i \tilde{A}_i^T A_i c = 0$$

Thus we may extend the n columns of A (and $\tilde{A}$) to a basis of $\mathbb{R}^{nm}$ and thereby form nonsingular nm matrices

$$C = \begin{bmatrix} \sqrt{\gamma_0} A_0 & C_0 \\ \vdots & \vdots \\ \sqrt{\gamma_{m-1}} A_{m-1} & C_{m-1} \end{bmatrix}$$

and

$$\tilde{C} = \begin{bmatrix} \sqrt{\gamma_0} \tilde{A}_0 & \tilde{C}_0 \\ \vdots & \vdots \\ \sqrt{\gamma_{m-1}} \tilde{A}_{m-1} & \tilde{C}_{m-1} \end{bmatrix}$$

where C_i, $\tilde{C}_i$, $i = 0, 1, \ldots, m - 1$ are $n \times n(m - 1)$ rectangular matrices.

To obtain matrices R and $\tilde{R}$ which satisfy equation (5.10) we transform C and $\tilde{C}$ into matrices R and $\tilde{R}$, respectively, of the form

$$C = RS, \quad \tilde{C} = \tilde{R}\tilde{S} \tag{5.19}$$

where S and $\tilde{S}$ are $nm \times nm$ nonsingular matrices

$$S = \begin{bmatrix} I_n & S_0 \\ O_n & S_1 \\ \vdots & \vdots \\ O_n & S_{m-1} \end{bmatrix}, \quad \tilde{S} = \begin{bmatrix} I_n & \tilde{S}_0 \\ O_n & \tilde{S}_1 \\ \vdots & \vdots \\ O_n & \tilde{S}_{m-1} \end{bmatrix}, \tag{5.20}$$

$S_i, \tilde{S}_i, i = 0, 1, \ldots, m - 1$ are $n \times (m - 1)n$ rectangular matrices and

$$C^T \tilde{C} = S^T \tilde{S}.$$

It would follow that (5.10) holds and R, $\tilde{R}$ have the desired form (5.8) and (5.9).

One way to transform C and $\tilde{C}$ into this desired form is by using the following easily verified variation of the Gram-Schmidt process.

LEMMA 5.4 *Let* $\{u_1, \ldots, u_n, f_1, \ldots, f_k\}$ *and* $\{u_1^*, \ldots, u_n^*, g_1, \ldots, g_k\}$ *be two bases for the space* $\mathbb{R}^{n+k}$ *with the property that the vectors* $\{u_1, \ldots, u_n\}$ *and* $\{u_1^*, \ldots, u_n^*\}$ *are a biorthogonal pair. Then there is a permutation* $\{p_1, \ldots, p_k\}$ *of* $\{1, 2, \ldots, k\}$ *such that sets of vectors* $\{u_1, \ldots, u_{n+k}\}$ *and* $\{u_1^*, \ldots, u_{n+k}^*\}$ *defined recursively as*

$$u_{n+i} = -\sum_{j=1}^{n+i-1} \langle u_j^*, f_i \rangle u_j + f_i$$

$$u_{n+i}^* = -\sum_{j=1}^{n+i-1} \frac{\langle g_{p_i}, u_j \rangle}{\langle g_{p_i}, u_{n+i} \rangle} u_j^* + \frac{g_{p_i}}{\langle g_{p_i}, u_{n+i} \rangle}$$

$i = 1, 2, \ldots, k$ *are a biorthogonal pair.*

Using this lemma it easily follows that we can choose S, $\tilde{S}$ as in (5.19) and (5.20) in the following form:

$$S = \begin{bmatrix} I_n & S_{0,0} & \cdots & S_{0,m-2} \\ O_n & S_{1,0} & \cdots & S_{1,m-2} \\ \vdots & \ddots & \ddots & \vdots \\ O_n & \cdots & O_n & S_{m-1,m-2} \end{bmatrix}$$

and

$$\tilde{S} = \begin{bmatrix} I_n & \tilde{S}_{0,0} & \cdots & \tilde{S}_{0,m-2} \\ O_n & \tilde{S}_{1,0} & \cdots & \tilde{S}_{1,m-2} \\ \vdots & \ddots & \ddots & \vdots \\ O_n & \cdots & O_n & \tilde{S}_{m-1,m-2} \end{bmatrix} \begin{bmatrix} I_n & O_{n \times (m-1)n} \\ O_{(m-1)n \times n} & P \end{bmatrix}$$

where $S_{i,j}, \tilde{S}_{i,j}, i = 0, 1, \ldots, m - 1, j = 0, 1, \ldots, m - 2$ are $n \times n$ matrices and P is an $(m - 1)n \times (m - 1)n$ permutation matrix. That is, up to the permutation matrix P both S and $\tilde{S}$ are upper triangular.

We remark that when A_0 and $\tilde{A}_0$ are *nonsingular* it is easy to obtain the nonsingular extension matrices C and $\tilde{C}$ described above. In fact, in this case we may choose C and $\tilde{C}$

to be the following nonsingular $(mn) \times (mn)$ matrices

$$C = \begin{bmatrix} \sqrt{\gamma_0}A_0 & O & \cdots & O & O \\ \sqrt{\gamma_1}A_1 & \sqrt{\gamma_1}I_n & \ddots & \vdots & \vdots \\ \sqrt{\gamma_2}A_2 & O & \ddots & O & O \\ \vdots & \vdots & \ddots & \sqrt{\gamma_{m-2}}I_n & O \\ \sqrt{\gamma_{m-1}}A_{m-1} & O & \cdots & O & \sqrt{\gamma_{m-1}}I_n \end{bmatrix}$$

and

$$\tilde{C} = \begin{bmatrix} \sqrt{\gamma_0}\tilde{A}_0 & O & \cdots & O & O \\ \sqrt{\gamma_1}\tilde{A}_1 & \sqrt{\gamma_1}I_n & \ddots & \vdots & \vdots \\ \sqrt{\gamma_2}\tilde{A}_2 & O & \ddots & O & O \\ \vdots & \vdots & \ddots & \sqrt{\gamma_{m-2}}I_n & O \\ \sqrt{\gamma_{m-1}}\tilde{A}_{m-1} & O & \cdots & O & \sqrt{\gamma_{m-1}}I_n \end{bmatrix}.$$

(A similar construction holds if *some* A_i, $i = 0, 1, \ldots, m-1$ is nonsingular.)

In the important special case $m = 2$ and under additional conditions on the matrices A_i, $\tilde{A}_i$, $i = 0, 1$ alternate particular solutions of the equations (5.4)–(5.6) can be given. For instance, when A_i, $\tilde{A}_i$, $i = 0, 1$ are *both* nonsingular then

$$B_0 = \gamma_1 \tilde{A}_0^{-T} G, \quad \tilde{B}_0 = \gamma_1 A_0^{-T} H$$

$$B_1 = -\gamma_0 \tilde{A}_1^{-T} G, \quad \tilde{B}_1 = -\gamma_0 A_1^{-T} H$$

where G, H are *any* $n \times n$ matrices which satisfy the condition

$$\gamma_0^{-1}\gamma_1^{-1} A_1^T \tilde{A}_1 A_0^T \tilde{A}_0 = HG^T$$

are solutions to equations (5.4)–(5.6).

6. Reconstruction and Decomposition Algorithms

In this section, we develop reconstruction and decomposition algorithms for biorthogonal wavelets in terms of refinement matrices A_i and $\tilde{A}_i$, and wavelet matrices B_i and $\tilde{B}_i$, $i = 0, 1, \ldots, m-1$. These algorithms may have applications to image compression and the numerical solution of boundary integral equations.

We write any integers $\ell, \ell' \in \{0, 1, \ldots, m^k - 1\}$ uniquely as

$$\ell = i_1 + mi_2 + \cdots + m^{k-1}i_k$$

and

$$\ell' = i'_1 + mi'_2 + \cdots + m^{k-1}i'_k$$

where $i_1, \ldots, i_k, i'_1, \ldots, i'_k \in \{0, 1, \ldots, m-1\}$. In what follows the numbers ℓ and ℓ' will be used interchangedly with their digits $i_1, \ldots, i_k$ and $i'_i, \ldots, i'_k$, relative the base m. Define vector fields

$$\mathbf{w}_{0,0}(t) := \mathbf{w}(t), \tag{6.1}$$

and

$$\mathbf{w}_{k,\ell}(t) := ((T_{i_k} \cdots T_{i_1})\mathbf{w}_{0,0})(t), \quad t \in E, \quad \ell = 0, 1, \ldots, m^k - 1, \tag{6.2}$$

where $\mathbf{w}(t)$ is the initial wavelet vector field mapping $E \to \mathbb{R}^{(m-1)n}$ and $T_i, i = 0, 1, \ldots, m-1$ are the linear operators as defined in (2.13), i.e.,

$$T_i = \sum_{j=0}^{m-1} q_{ij} P_j, \quad i \in \{0, 1, \ldots, m-1\}. \tag{6.3}$$

Clearly, the vector fields $\mathbf{w}_{k,\ell}$ also map $E \to \mathbb{R}^{(m-1)n}$ and by Theorem 4.6 the components of $\mathbf{w}_{k,\ell}, \ell = 0, 1, \ldots, m^k - 1$ form a basis for the space W_k.

LEMMA 6.1 *For a positive integer k and $\ell, \ell' \in \{0, 1, \cdots, m-1\}$, there holds the formula*

$$\mathbf{w}_{k,\ell}((\phi_{i'_k} \circ \cdots \circ \phi_{i'_1})(t)) = \frac{q_{i_k i'_k} q_{i_{k-1} i'_{k-1}} \cdots q_{i_1 i'_1} \mathbf{w}_{0,0}(t)}{\sqrt{\gamma_{i'_1} \cdots \gamma_{i'_k}}}, \quad t \in E. \tag{6.4}$$

Proof: Notice that

$$G_i T_j = q_{ji} I, \quad i, j = 0, 1, \ldots, m-1. \tag{6.5}$$

Applying $G_{i'_k}$ to both sides of equation (6.2) yields

$$(G_{i'_k}\mathbf{w}_{k,\ell})(t) = (G_{i'_k}(T_{i_k} \cdots T_{i_1})\mathbf{w}_{0,0})(t).$$

By the definition of $G_{i'_k}$ and equation (6.5), we have

$$\sqrt{\gamma_{i'_k}}\mathbf{w}_{k,\ell}(\phi_{i'_k}(t)) = q_{i_k i'_k}((T_{i_{k-1}} \cdots T_{i_1})\mathbf{w}_{0,0})(t).$$

Repeatedly applying this procedure leads to equation (6.4). ∎

For $\ell, \ell' \in \{0, 1, \ldots, m^k - 1\}$, we define an $m^k \times m^k$ matrix V by

$$V := (v_{\ell\ell'})_{\ell,\ell'=0,1,\ldots,m^k-1} \tag{6.6}$$

where

$$v_{\ell \ell'} := q_{i_k i_k'} \cdots q_{i_1 i_1'}.$$

Note that V is an orthogonal matrix. Thus, equation (6.4) can be written as

$$\sqrt{\gamma_{i_1'} \cdots \gamma_{i_k'}} \mathbf{w}_{k,\ell}((\phi_{i_k'} \circ \cdots \circ \phi_{i_1'})(t)) = v_{\ell \ell'} \mathbf{w}_{0,0}(t). \tag{6.7}$$

Let us now consider the matrix interpretation of the biorthogonal wavelet transform between successive levels. We begin with the basic formulas

$$F_{k+1} = F_k \oplus W_k, \tag{6.8}$$

and

$$\tilde{F}_{k+1} = \tilde{F}_k \oplus \tilde{W}_k, \tag{6.9}$$

From equation (6.8), for $g \in F_{k+1}$, there exist $g_0 \in F_k$ and $g_1 \in W_k$ such that

$$g(t) = g_0(t) + g_1(t), \quad t \in E. \tag{6.10}$$

Since $g_0 \in F_k$, there exists a vector $\mathbf{s}_{k,\ell} \in \mathbb{R}^n$ such that

$$\sqrt{\gamma_{i_1} \cdots \gamma_{i_k}} g_0((\phi_{i_k} \circ \cdots \circ \phi_{i_1})(t)) = \mathbf{s}_{k,\ell}^T \mathbf{f}(t). \tag{6.11}$$

Note that W_k is spanned by the components of $\mathbf{w}_{k,\ell}$, $\ell = 0, 1, \ldots, m^k - 1$. Since $g_1 \in W_k$, there exist vectors $\mathbf{t}_{k,\ell} \in \mathbb{R}^{(m-1)n}$, $\ell = 0, 1, \ldots, m^k - 1$, such that

$$g_1(t) = \sum_{\ell=0}^{m^k-1} \mathbf{t}_{k,\ell}^T \mathbf{w}_{k,\ell}(t). \tag{6.12}$$

By Lemma 6.1, we conclude that for $\ell' = 0, 1, \ldots, m^k - 1$,

$$\begin{aligned} g_1((\phi_{i_k'} \circ \cdots \circ \phi_{i_1'})(t)) &= \sum_{\ell=0}^{m^k-1} \mathbf{t}_{k,\ell}^T \mathbf{w}_{k,\ell}((\phi_{i_k'} \circ \cdots \circ \phi_{i_1'})(t)) \\ &= \sum_{\ell=0}^{m^k-1} \mathbf{t}_{k,\ell}^T \frac{v_{\ell,\ell'}}{\sqrt{\gamma_{i_1'} \cdots \gamma_{i_k'}}} \mathbf{w}_{0,0}(t), \quad 0 \le t \le 1. \end{aligned} \tag{6.13}$$

On the other hand, since $g \in F_{k+1}$, there exist vectors $\mathbf{s}_{k+1,\ell} \in \mathbb{R}^n$, $\ell = 0, 1, \ldots, m^{k+1} - 1$, such that

$$\sqrt{\gamma_{i_1'} \cdots \gamma_{i_{k+1}'}} g((\phi_{i_{k+1}'} \circ \cdots \circ \phi_{i_1'})(t)) = \mathbf{s}_{k+1,\ell}^T \mathbf{f}(t). \tag{6.14}$$

Let

$$\hat{\ell} := i_2' + m i_3' + \cdots + m^{k-1} i_{k+1}'.$$

Then,

$$\ell' = m\hat{\ell} + i_1'.$$

By equations (6.10)–(6.13), we have

$$
\begin{aligned}
&g((\phi_{i'_{k+1}} \circ \cdots \circ \phi_{i'_1})(t)) \\
&= g_0((\phi_{i'_{k+1}} \circ \cdots \circ \phi_{i'_1})(t)) + g_1((\phi_{i'_{k+1}} \circ \cdots \circ \phi_{i'_1})(t)) \\
&= g_0((\phi_{i'_{k+1}} \circ \cdots \circ \phi_{i'_2})(\phi_{i'_1}(t))) + g_1((\phi_{i'_{k+1}} \circ \cdots \circ \phi_{i'_2})(\phi_{i'_1}(t))) \\
&= \frac{1}{\sqrt{\gamma_{i'_1} \cdots \gamma_{i'_{k+1}}}} \left[\mathbf{s}_{k,\hat{\ell}}^T \sqrt{\gamma_{i'_1}} \mathbf{f}(\phi_{i'_1}(t)) + \sum_{\ell=0}^{m^k-1} \mathbf{t}_{k,\ell}^T v_{\ell,\hat{\ell}} \sqrt{\gamma_{i'_1}} \mathbf{w}_{0,0}(\phi_{i'_1}(t)) \right] \\
&= \frac{1}{\sqrt{\gamma_{i'_1} \cdots \gamma_{i'_{k+1}}}} \left[\mathbf{s}_{k,\hat{\ell}}^T \sqrt{\gamma_{i'_1}} A_{i'_1}^T \mathbf{f}(t) + \sum_{\ell=0}^{m^k-1} \mathbf{t}_{k,\ell}^T v_{\ell,\hat{\ell}} \sqrt{\gamma_{i'_1}} B_{i'_1}^T \mathbf{f}(t) \right].
\end{aligned}
$$

It follows from this equation and equation (6.14) that

$$\mathbf{s}_{k+1,m\hat{\ell}+i_1'} = \sqrt{\gamma_{i'_1}} A_{i'_1} \mathbf{s}_{k,\hat{\ell}} + \sqrt{\gamma_{i'_1}} B_{i'_1} \sum_{\ell=0}^{m^k-1} \mathbf{t}_{k,\ell} v_{\ell,\hat{\ell}},$$

where

$$i_1' \in \{0, 1, \ldots, m-1\}.$$

Therefore, we have the decomposition algorithm

$$
\begin{pmatrix} \mathbf{s}_{k+1,m\hat{\ell}} \\ \vdots \\ \mathbf{s}_{k+1,m\hat{\ell}+m-1} \end{pmatrix}
= \begin{pmatrix} \sqrt{\gamma_0} A_0 & \sqrt{\gamma_0} B_0 \\ \vdots & \vdots \\ \sqrt{\gamma_{m-1}} A_{m-1} & \sqrt{\gamma_{m-1}} B_{m-1} \end{pmatrix}
\begin{pmatrix} \mathbf{s}_{k,\hat{\ell}} \\ \sum_{\ell=0}^{m^k-1} v_{\ell,\hat{\ell}} \mathbf{t}_{k,\ell} \end{pmatrix},
$$

and in view of equations (5.8)–(5.10) the reconstruction algorithm

$$
\begin{pmatrix} \mathbf{s}_{k,\hat{\ell}} \\ \sum_{\ell=0}^{m^k-1} v_{\ell,\hat{\ell}} \mathbf{t}_{k,\ell} \end{pmatrix}
= \begin{pmatrix} \sqrt{\gamma_0} \tilde{A}_0^T & \cdots & \sqrt{\gamma_{m-1}} \tilde{A}_{m-1}^T \\ \sqrt{\gamma_0} \tilde{B}_0^T & \cdots & \sqrt{\gamma_{m-1}} \tilde{B}_{m-1}^T \end{pmatrix}
\begin{pmatrix} \mathbf{s}_{k+1,m\hat{\ell}} \\ \vdots \\ \mathbf{s}_{k+1,m\hat{\ell}+m-1} \end{pmatrix}.
$$

Similarly, we can establish the decomposition and reconstruction algorithms from equation (6.9) for the space $\tilde{F}_{k+1}$.

Some of the results in this section appear in preliminary form in [19].

7. Special Cases

In this section we will present some additional information about the mappings ϕ_i, $i = 0, 1, \ldots, m-1$. Most of our remarks are for the case of the interval $[0, 1]$. However, we begin by recalling some relevant information contained in the paper [17]. For manifolds E constructions of smooth mappings ϕ_i, $i = 0, 1, \ldots, m - 1$ are provided by the discussions in this paper. To review some aspects of [17], we suppose

$$\mu(x) > 0, \quad x \in E,$$

let

$$\gamma_i := \int_{E_i} \mu(x)dx, \quad i = 0, 1, \ldots, m - 1$$

and consider the problem of finding bijective mappings $\phi_i : E \to E_i$, $i = 0, 1, \ldots, m - 1$ such that the linear operators

$$(P_i f)(x) = \begin{cases} \frac{1}{\sqrt{\gamma_i}} f(\phi_i^{-1}(x)), & x \in E_i \\ 0, & x \notin E_i \end{cases}$$

$i = 0, 1, \ldots, m - 1$, are isometries on $L^2(E, \rho)$, where $d\rho(x) := \mu(x)dx$. When ϕ_i, $i = 0, 1, \ldots, m - 1$ are differentiable the conditions which insure that these mappings ϕ_i, $i = 0, 1, \ldots, m - 1$ have this property are given by the equation

$$\gamma_i \mu(x) = \mu(\phi_i(x))J_{\phi_i}(x), \quad x \in E, \quad i = 0, 1, \ldots, m - 1.$$

As used earlier J_{ϕ_i} denotes the Jacobian of the mapping ϕ_i.

To find the mappings ϕ_i, $i = 0, 1, \ldots, m - 1$, we consider the following problem: given a positive measurable function μ on the set E find a bijective differentiable map $\alpha_\mu : E \to E$ whose Jacobian has the value μ. That is, $J_{\alpha_\mu}(x) = \mu(x)$, $x \in E$ and $\alpha_\mu : E \to E$, bijectively. If we assume the set E_i is an affine image of E, that is, $\psi_i : E \to E_i$ bijectively, ψ_i affine, then we can obtain ϕ_i in the following way. We define the weight function

$$\tilde{\mu}_i(x) := \frac{c_i}{\gamma_i} \mu(\psi_i(x)), \quad x \in E, \quad i = 0, 1, \ldots, m - 1$$

where

$$c_i = |J_{\psi_i}| = J_{\psi_i}, \quad i = 0, 1, \ldots, m - 1$$

(so that ψ_i is orientation preserving). Then

$$\int_E \tilde{\mu}_i(x)dx = 1$$

and the mapping

$$\hat{\alpha}_i := \alpha_{\tilde{\mu}_i} \circ \psi_i^{-1}, \quad i = 0, 1, \ldots, m - 1$$

is a bijective mapping $\hat{\alpha}_i : E_i \to E$ with the property that

$$J_{\hat{\alpha}_i}(x) = \frac{\mu(x)}{\gamma_i}, \quad x \in E.$$

Hence the mapping

$$\phi_i := \hat{\alpha}_i^{-1} \circ \alpha_\mu = \psi_i \circ \alpha_{\hat{\mu}_i}^{-1} \circ \alpha_\mu, \quad i = 0, 1, \dots, m-1$$

is likewise a bijection $\phi_i : E \to E_i$ such that

$$\hat{\alpha}_i \circ \phi_i = \alpha_\mu.$$

Differentiating this equation it follows that for $x \in E$ and $i = 0, 1, \dots, m-1$

$$\mu(x) = J_{\alpha_\mu}(x) = J_{\hat{\alpha}_i}(\phi_i(x))J_{\phi_i}(x) = \frac{1}{\gamma_i}\mu(\phi_i(x))J_{\phi_i}(x).$$

That is, ϕ_i is the desired mapping. We record below an explicit form of the mapping α_μ for a square which is suggested by [17].

LEMMA 7.1 *Let $\mu(x, y)$ be a positive Lebesgue measurable function on the square $S :=$ $[0, 1] \times [0, 1]$ such that*

$$\int_S \mu(x, y)dxdy = 1.$$

Then the mapping $\alpha_\mu : S \to S$ given by

$$\alpha_\mu(x, y) = (u(x, y), v(x, y)), \quad (x, y)^T \in S,$$

where

$$u(x, y) = \int_0^x \left(\int_0^1 \mu(r, s)ds \right) dr, \quad (x, y)^T \in S,$$

$$v(x, y) = \frac{\int_0^y \mu(x, s)ds}{\int_0^1 \mu(x, s)ds}, \quad (x, y)^T \in S$$

is one to one and onto S with Jacobian μ.

Proof: A direct computation verifies that $J_{\alpha_\mu} = \mu$. Note that $u(x, y)$ is independent of y and in x maps $[0, 1]$ one to one and onto $[0, 1]$. Similarly, for each x, $v(x, \cdot)$ is a one to one and onto mapping of $[0, 1]$. Hence α_μ has all the desired properties.   ∎

As a consequence we obtain the following fact.

PROPOSITION 7.2 *Let μ satisfy the hypotheses of Lemma 7.1. Define the affine mappings*

$$\psi_e(x) = \frac{1}{2}x + e, \quad e \in ext\ S$$

and corresponding sets $E_e := \psi_e(S)$, $e \in ext\ S$. Then the mappings

$$\phi_e = \psi_e \circ \alpha_{\Sigma_e}^{-1} \circ \alpha_\mu$$

where

$$\Sigma_e(x) := \frac{1}{4\gamma_e}\mu(\psi_e(x)), \quad x \in S$$

$$\gamma_e := \int_{E_e} \mu(x)dx$$

yield isometries

$$(P_e f)(x) = \begin{cases} \frac{1}{\sqrt{\gamma_e}} f(\phi_e^{-1}(x)), & x \in E_e \\ 0, & x \notin E_e \end{cases}$$

such that

$$P_e P_{e'}^* = \delta_{ee'}, \quad e, e' \in ext\ S.$$

Our next example studies the case when $E = [0, 1]$ and $E_i = [\frac{i}{m}, \frac{i+1}{m}]$, $i = 0, 1, \ldots, m - 1$, $m \geq 2$. In this case, the mappings are given in the next lemma.

LEMMA 7.3 *Suppose that ρ is a strictly increasing bijection of $[0, 1]$. The strictly increasing bijection ϕ_i that maps $[0, 1]$ onto $[\frac{i}{m}, \frac{i+1}{m}]$, so that P_i, $i = 0, 1, \ldots, m - 1$, defined in (2.3) are isometries on $L^2([0, 1], \rho)$ are given by*

$$\phi_i(x) = \rho^{-1}\left(\gamma_i \rho(x) + \rho\left(\frac{i}{m}\right)\right), \quad x \in [0, 1]$$

and

$$\gamma_i = \rho\left(\frac{i+1}{m}\right) - \rho\left(\frac{i}{m}\right)$$

for $i = 0, 1, \ldots, m - 1$. Moreover, the inverse of ϕ_i is given by

$$\phi_i^{-1}(x) = \rho^{-1}\left(\frac{1}{\gamma_i}\left(\rho(x) - \rho\left(\frac{i}{m}\right)\right)\right), \quad x \in \left[\frac{i}{m}, \frac{i+1}{m}\right].$$

Proof: A direct computation verifying that

$$\frac{1}{\gamma_i} \rho(\phi_i(x)) - \rho(x) = \text{constant}, \quad x \in [0, 1]$$

yields the validity of this result.   ∎

In what follows we provide an *alternate* construction of the functions ϕ_i, $i = 0, 1, \ldots, m-1$, in Lemma 7.3 in terms of a specific refinable function σ that *only* depends on the constants, γ_i, $i = 0, 1, \ldots, m - 1$. To this end, we consider the 2×2 matrices

$$R_j := \begin{pmatrix} 1 - \sum_{i=0}^{j-1} \gamma_i & \sum_{i=0}^{j-1} \gamma_i \\ 1 - \sum_{i=0}^{j} \gamma_i & \sum_{i=0}^{j} \gamma_i \end{pmatrix}, \quad j = 0, 1, \ldots, m - 1,$$

(a vacuous sum is set to zero). There exists a continuous strictly increasing bijection σ of $[0, 1]$ such that the planar curve

$$\mathbf{r}(t) := \begin{pmatrix} 1 - \sigma(t) \\ \sigma(t) \end{pmatrix}$$

satisfies the refinement equation

$$\mathbf{r}\left(\frac{t + j}{m}\right) = R_j^T \mathbf{r}(t), \quad t \in [0, 1], \quad j = 0, 1, \ldots, m - 1. \tag{7.1}$$

To prove the existence of σ we follow [18] and [16] and compute the eigenvector of R_i^T corresponding to eigenvalue one. For this purpose, we introduce the constants

$$\sigma_j = \sum_{i=0}^{j} \gamma_i, \quad j = 0, 1, \ldots, m - 1,$$

where we also set $\sigma_{-1} = 0$ and note that $\sigma_{m-1} = 1$. By a straightforward computation, we see that

$$u_j = \left(\frac{1 - \sigma_j}{1 + \sigma_{j-1} - \sigma_j}, \frac{\sigma_{j-1}}{1 + \sigma_{j-1} - \sigma_j}\right)^T, \quad j = 0, 1, \ldots, m - 1$$

is an eigenvector of the matrix R_j^T, $j = 0, 1, \ldots, m - 1$ associated with the eigenvalue one. Moreover, since $u_0 = (1, 0)^T$ and $u_{m-1} = (0, 1)^T$ it follows that

$$R_j^T u_0 = R_{j-1}^T u_{m-1}, \quad j = 1, 2, \ldots, m.$$

Therefore, since R_j, $j = 0, 1, \ldots, m - 1$ are stochastic matrices with positive columns, by [18] and [16], the existence of σ is assured.

Another way to express formula (7.1) is by considering the function

$$\phi(t) = \begin{cases} \sigma(t), & 0 \le t \le 1 \\ 1 - \sigma(t-1), & 1 \le t \le 2 \\ 0, & t \notin [0, 2]. \end{cases}$$

The fact that σ is increasing follows the same argument used in [16, p. 20–25] for the case $m = 1$. Additionally, it follows from the refinement equations (7.1) that the function σ satisfies the equations

$$\gamma_j \sigma(t) + \sigma_{j-1} = \sigma\left(\frac{t+j}{m}\right), \quad t \in [0, 1], \quad j = 0, 1, \ldots, m-1. \tag{7.2}$$

Therefore ϕ satisfies the refinement equation

$$\phi(t) = \sum_{j=0}^{2m-2} \mu_j \phi(mt - j), \quad t \in \mathbb{R}$$

where we set

$$\mu_j := \sigma_j, \quad \mu_{m+j} := 1 - \sigma_j \quad j = 0, 1, \ldots, m-1.$$

Next, we define a real valued function H on $[0, 1]$ by the formula

$$\rho(x) = \sigma(H(x)), \quad x \in [0, 1]. \tag{7.3}$$

LEMMA 7.4 *Suppose ρ is continuous and strictly increasing on $[0, 1]$. Then the function H is a strictly increasing continuous function on $[0, 1]$ and H maps $[\frac{j}{m}, \frac{j+1}{m}]$ bijectively $[\frac{j}{m}, \frac{j+1}{m}]$ for $j = 0, 1, \ldots, m-1$.*

Proof: Since both ρ and σ are strictly increasing and continuous on $[0, 1]$, it follows that H is likewise strictly increasing and continuous on $[0, 1]$.
  To prove that H maps $[\frac{j}{m}, \frac{j+1}{m}]$ onto $[\frac{j}{m}, \frac{j+1}{m}]$ it suffices to show that

$$H\left(\frac{j}{m}\right) = \frac{j}{m}, \quad j = 0, 1, \ldots, m.$$

In (7.2) letting $t = 0$, we obtain that

$$\sigma\left(\frac{j}{m}\right) = \gamma_0 + \cdots + \gamma_{j-1} = \sigma_{j-1}, \quad j = 0, 1, \ldots, m-1.$$

On the other hand, we have

$$\sigma_{j-1} = \sum_{i=0}^{j-1} \gamma_i = \sum_{i=0}^{j-1} \left(\rho\left(\frac{i+1}{m}\right) - \rho\left(\frac{i}{m}\right)\right)$$

$$= \rho\left(\frac{j}{m}\right) = \sigma\left(H\left(\frac{j}{m}\right)\right) \quad j = 0, 1, \ldots, m-1.$$

This implies that

$$\sigma\left(\frac{j}{m}\right) = \sigma\left(H\left(\frac{j}{m}\right)\right) \quad j = 0, 1, \ldots, m-1.$$

Since σ is bijection, we conclude that

$$H(\frac{j}{m}) = \frac{j}{m}, \quad j = 0, 1, \ldots, m-1. \qquad \blacksquare$$

The following theorem gives an expression for the function ϕ_i and the solution of refinement equations.

THEOREM 7.5 *Let ϕ_i be the functions given in Lemma 7.3. Then*

$$\phi_j(x) = H^{-1}\left(\frac{H(x)+j}{m}\right), \quad j = 0, 1, \ldots, m-1.$$

Furthermore, assume

$$\hat{\mathbf{f}} = (\hat{f}_1, \ldots, \hat{f}_n) : L^2([0, 1], \rho) \to \mathbb{R}^n$$

satisfies the refinement equation

$$\hat{\mathbf{f}}\left(\frac{t+j}{m}\right) = A_i^T \hat{\mathbf{f}}(t), \quad i = 0, 1, \ldots, m-1. \qquad (7.4)$$

Then the curve $\mathbf{f}$

$$\mathbf{f}(t) := \hat{\mathbf{f}}(H(t)), \quad t \in [0, 1]$$

satisfies the refinement equations

$$\mathbf{f}(\phi_j(t)) = A_j^T \mathbf{f}(t), \quad t \in [0, 1], \quad j = 0, 1, \ldots, m-1.$$

Proof: It follows from Lemma 7.3 and equations (7.2) and (7.3) that for $j = 0, 1, \ldots, m-1$,

$$\begin{aligned} \phi_j(x) &= \rho^{-1}(\gamma_j \sigma(H(x)) + \gamma_0 + \cdots + \gamma_{j-1}) \\ &= \rho^{-1}\left(\sigma\left(\frac{H(x)+j}{m}\right)\right) \\ &= H^{-1}\left(\frac{H(x)+j}{m}\right). \end{aligned}$$

It remains to prove the second part of this theorem. It follows from the first part of this theorem that

$$H(\phi_j(t)) = \frac{H(t)+j}{m}, \quad j = 0, 1, \ldots, m-1.$$

By the definition of **f** and equation (7.4), we have

$$
\begin{aligned}
\mathbf{f}(\phi_j(t)) &= \hat{\mathbf{f}}(H(\phi_j(t))) \\
&= \hat{\mathbf{f}}\left(\frac{H(t)+j}{m}\right) \\
&= A_j^T \hat{\mathbf{f}}(H(t)) \\
&= A_j^T \mathbf{f}(t), \quad t \in [0, 1].
\end{aligned}
$$

∎

Our final example is in fact the one which provided the initial motivation for this paper. It concerns the *Chebyshev* weight function on the interval $[-1, 1]$. Thus we consider the function

$$
\rho_c(x) := -\frac{1}{\pi} \arccos x + 1, \quad x \in [-1, 1]. \tag{7.5}
$$

In this case

$$
\rho_c'(x) = \frac{1}{\pi} \frac{dx}{\sqrt{1-x^2}}, \quad x \in (-1, 1) \tag{7.6}
$$

and for this weight function we partition the interval according to the location of the *extrema* of the m-th Chebyshev polynomial

$$
T_m(x) = \cos m\theta, \quad x = \cos\theta. \tag{7.7}
$$

That is, we consider the nonuniform partition of $[-1, 1]$ using the intervals

$$
E_i := [x_{i+1}, x_i], \quad i = 0, 1, \ldots, m-1 \tag{7.8}
$$

where

$$
x_i := \cos\frac{i\pi}{m}, \quad i = 0, 1, \ldots, m. \tag{7.9}
$$

To analyze this case we need to make two improvements in the statements made in Lemma 7.3. First, we need to consider nonuniform partitions and second allows for the possibility of using mapping ϕ_i which are *decreasing*. Specifically, given any interval $[a, b]$, some subinterval $[c, d]$ of it and an increasing function ρ such that $\rho(a) = 0$, $\rho(b) = 1$ so that

$$
\int_a^b d\rho(x) = 1
$$

then either the *increasing* mapping

$$
\phi(x) = \rho^{-1}(\gamma\rho(x) + \rho(c)), \quad x \in [a, b] \tag{7.10}
$$

$$
\gamma := \rho(d) - \rho(c) \tag{7.11}
$$

or the *decreasing* mapping

$$\phi(x) := \rho^{-1}(-\gamma\rho(x) + \rho(d)), \quad x \in [a, b] \tag{7.12}$$

has the property that $\phi : [a, b] \to [c, d]$, surjectively and

$$\frac{1}{\gamma} \int_c^d f(\phi^{-1}(x))d\rho(x) = \int_a^b f(x)d\rho(x), \quad f \in L^1([a, b], \rho). \tag{7.13}$$

Let us apply this observation to the Chebyshev weight ρ'_c. In this case we choose for the interval E_i given by (7.8) and (7.9) the increasing mapping when T_m increases on E_i and the decreasing mapping when T_m decreases. Let us denote these mappings by $\phi_0, \ldots, \phi_{m-1}$, and note that $\phi_i : [-1, 1] \to E_i, i = 0, 1, \ldots, m-1$. We claim that for $i = 0, 1, \ldots, m-1$

$$\phi_i^{-1}(x) = T_m(x), \quad x \in E_i. \tag{7.14}$$

The proof of this formula is provided only for the case that T_m increases on E_i. The other case follows similarly.

Generally, we observe that

$$\gamma_i := \rho_c(x_i) - \rho_c(x_{i+1}) = \frac{1}{m}, \quad i = 0, 1, \ldots, m-1$$

and for $i = 0, 1, \ldots, m-1$

$$\phi_i(x) = \begin{cases} \rho_c^{-1}(\gamma_i \rho_c(x) + \rho_c(x_{i+1})), & \text{if } T_m \text{ increases on } E_i \\ \rho_c^{-1}(-\gamma_i \rho_c(x) + \rho_c(x_i)), & \text{if } T_m \text{ decreases on } E_i \end{cases}$$

When T_m increases on E_i equation (7.14) means that

$$\rho_c(x) = \frac{1}{m}\rho_c(T_m(x)) - \frac{i+1}{m} + 1, \quad x \in E_i. \tag{7.15}$$

For $x = \cos\theta, \theta \in [\frac{i}{m}, \frac{i+1}{m}]$ the right hand side of equation (7.15) becomes

$$\frac{1}{m}\rho_c(\cos(m\theta)) - \frac{i+1}{m} + 1 = -\frac{\theta}{\pi} + 1$$

which is in agreement with the left hand side of equation (7.15). This verifies formula (7.14) and means, in view of our means by our general receipe (2.3) that the linear operators

$$P_i f = \sqrt{m}\chi_{E_i} f \circ T_m, \quad i = 0, 1, \ldots, m-1, \quad f \in L^2([-1, 1], \rho_c)$$

are a family of mutually orthogonal isometries on $L^2([-1, 1], \rho_c)$.

To complete this example we need to identify a suitable collection of refinable curves. To this end, we consider the collection of affine mappings $a_i, i = 0, 1, \ldots, m-1$ defined by

$$a_i(t) = \begin{cases} \frac{i+t}{m}, & i \text{ even}, i = 0, 1, \ldots, m-1 \\ \frac{i+1-t}{m}, & i \text{ odd}, i = 0, 1, \ldots, m-1. \end{cases}$$

Thus $a_i : [0, 1] \to [\frac{i}{m}, \frac{i+1}{m}]$, $i = 0, 1, \ldots, m - 1$ bijectively with a_i increasing for i even and a_i decreasing for i odd, $i = 0, 1, \ldots, m - 1$. In analog with formula (7.14) we note that for $i = 0, 1, \ldots, m - 1$,

$$a_i^{-1}(t) = a(t), \quad t \in \left[\frac{i}{m}, \frac{i+1}{m} \right]$$

where a is the piecewise continuous linear map of $[0, 1]$ onto $[0, 1]$ with breakpoints $\frac{i}{m}$, $i = 1, 2, \ldots, m - 1$ with the interpolatory property that

$$a\left(\frac{i}{m} \right) = \frac{1}{2}[-1 - (-1)^i], \quad i = 0, 1, \ldots, m - 1.$$

Since these mappings are affine there are clearly refinable curves $\hat{\mathbf{f}} : [0, 1] \to \mathbb{R}^n$ and corresponding $n \times n$ matrices $A_0, \ldots, A_{m-1}$ such that for $i = 0, 1, \ldots, m - 1$

$$\hat{\mathbf{f}}(a_i(t)) = A_i^T \hat{\mathbf{f}}(t), \quad t \in [0, 1] \tag{7.17}$$

$$\mathbf{v}^T \hat{\mathbf{f}}(t) = 1, \quad t \in [0, 1].$$

We can, in fact merely take $\hat{\mathbf{f}} = (\hat{f}_1, \ldots, \hat{f}_n)^T$ to be a polynomial curve of degree $\leq n - 1$, so that polynomials of degree $n - 1$, $\hat{f}_1, \ldots, \hat{f}_n$ are linearly independent on $[0, 1]$.

Let $\hat{\mathbf{f}}$ be *any* refinable curve satisfying equation (7.17) for some $n \times n$ matrices $A_0, \ldots, A_{m-1}$ and define the curve $\mathbf{f} : [-1, 1] \to \mathbb{R}^n$ by the formula

$$\mathbf{f}(t) := \hat{\mathbf{f}}\left(\frac{1}{\pi} \arccos t \right), \quad t \in [-1, 1]. \tag{7.18}$$

Equivalently, we have

$$\hat{\mathbf{f}}(t) = \mathbf{f}(\cos \pi t), \quad t \in [0, 1].$$

Substituting this formula into equation (7.17) gives that for $t \in [0, 1]$ and $i = 0, 1, \ldots, m - 1$

$$A_i^T \mathbf{f}(\cos \pi t) = \begin{cases} \mathbf{f}\left(\cos \pi \frac{i+t}{m} \right), & i \text{ even} \\ \mathbf{f}\left(\cos \pi \frac{i+1-t}{m} \right), & i \text{ odd.} \end{cases}$$

Equivalently we have for $y \in [\frac{i\pi}{m}, \frac{(i+1)\pi}{m}]$ and $i = 0, 1, \ldots, m - 1$ that

$$\mathbf{f}(\cos y) = \begin{cases} A_i^T \mathbf{f}(\cos(my - \pi i)), & i \text{ even} \\ A_i^T \mathbf{f}(\cos((i + 1)\pi - my), & i \text{ odd} \end{cases}$$
$$= A_i^T \mathbf{f}(\cos my), \quad i = 0, 1, \ldots, m - 1.$$

That is, for $i = 0, 1, \ldots, m - 1$

$$\mathbf{f}(x) = A_i^T \mathbf{f}(T_m(x)), \quad x \in E_i,$$

or, in view of formula (7.14)

$$\mathbf{f}(\phi_i(x)) = A_i^T \mathbf{f}(x), \quad x \in [-1, 1], \quad i = 0, 1, \ldots, m - 1.$$

So we have confirmed that the curve $\mathbf{f}$ given by (7.18) is indeed refinable relative to the mappings (7.14). Therefore the wavelet construction of Sections 1–6 applies in an explicit fashion for the Chebyshev weight!

Acknowledgements

This work is partially supported by NSF under grant DMS-9504780. The first author was partially supported by the Alexander von Humboldt Foundation. The second author was partially supported by NASA under grant NAG 3-1312 and NASA-OAI Summer Faculty Fellowship (1995).

References

1. C. A. Micchelli and Y. Xu, "Using the matrix refinement equation for the construction of wavelets on invariant sets," *Appl. Comp. Harmonic Anal.*, vol. 1, 1994, pp. 391–401.

2. C. A. Micchelli and Y. Xu, "Using the matrix refinement equation for the construction of wavelets II: Smooth wavelets on [0, 1]," in *Approximation and Computation, A Festschrift in Honor of Walter Gautschi*, R. Zahar (ed.), Birkhauser, Boston, 1995, pp. 435–457.

3. B. K. Alpert, "A class of bases in L^2 for sparse representation of integral operators," *SIAM J. Math. Anal.*, vol. 24, 1993, pp. 246–262.

4. A. Alpert, G. Beylkin, R. Coifman, and V. Rokhlin, "Wavelet-like bases for the fast solution of second-kind integral equations," *SIAM J. Sci. Comput.*, 14, 1993, pp. 159–184.

5. G. Beylkin, R. Coifman, and V. Rokhlin, "Fast wavelet transforms and numerical algorithms I," *Comm. Pure Appl. Math.*, vol. XLIV, 1991, pp. 141–183.

6. W. Dahmen, S. Proessdorf, and R. Schneider, "Wavelet approximation methods for pseudodifferential equations I: Stability and convergence," *Math. Z.*, vol. 215, 1994, pp. 583–620.

7. W. Dahmen, S. Proessdorf, and R. Schneider, "Wavelet approximation methods for pseudodifferential equations II: Matrix compression and fast solution," *Advances in Computational Mathematics*, vol. 1, 1993, pp. 259–335.

8. R. Hu, C. A. Micchelli, and Y. Xu, "Weakly singular Fredholm integral equations II: Singularity preserving wavelet-collocation methods," *Proceedings of the 8th Texas International Conference on Approximation Theory*, submitted.

9. C. A. Micchelli and Y. Xu, "Weakly singular Fredholm integral equations I: singularity preserving wavelet-Galerkin methods," *Proceedings of the 8th Texas International Conference on Approximation Theory*, submitted.

10. C. A. Micchelli, Y. Xu and Y. Zhao, "Wavelet methods for multidimensional integral equations," in preparation.

11. T. von Petersdorff and C. Schwab, "Wavelet approximations for first kind boundary integral equations on polygons," *Numer. Math.*, to appear.

12. T. von Petersdorff, C. Schwab, and R. Schneider, "Multiwavelets for second kind integral equations," preprint, 1995.

13. H. L. Royden, *Real Analysis,* 1st edition, New York: Macmillan, 1963.

14. J. E. Hutchinson, "Fractals and self similarity," *Indiana Univ. Math. J.*, vol. 30, 1981, pp. 713–747.

15. K. Gröchenig and W. R. Madych, "Multiresolution analysis, Haar bases, and self-similar tiling of $\mathbb{R}^n$," *IEEE Transation on Information Theory*, vol. 38, 1992, pp. 556–568.

16. C. A. Micchelli, *Geometric Modeling*, CBMS Vol. 67, Philidelphia: SIAM Publications, 1994.

17. J. Moser, "On the volume elements on a manifold," *Trans. A.M.S.*, vol. 120, 1965, pp. 286–294.

18. C. A. Micchelli and H. Prautzsch, "Uniform refinement of curves," *Linear Algebra Appl.*, vol. 114/115, 1989, pp. 841–870.

19. C. A. Micchelli and Y. Xu, "Wavelets on a finite interval: Decomposition and reconstruction algorithms," in *Proceedings of the Workshop on Wavelets*, National University of Singapore, June 29–July 1, 1994, Singapore.

20. G. Szego, *Orthogonal Polynomials*, Amer. Math. Soc. Coll. Pub. XXII, 1939.

Multidimensional Systems and Signal Processing, 8, 71–88 (1997)
© 1997 Kluwer Academic Publishers, Boston.

Zero-Phase Filter Bank and Wavelet Code r Matrices: Properties, Triangular Decompositions, and a Fast Algorithm

T. G. MARSHALL, JR.*   marshall@alvand.rutgers.edu
Dept. of ECE, Rutgers University, Piscataway, New Jersey 08854

Abstract. The class of digital filter banks (DFB's) and wavelets composed of zero-phase filters is particularly useful for image processing because of the desirable filter responses and the possibility of using the McClellan transformation for two-dimensional design. In this paper unimodular polyphase matrices with the identity matrix for Smith canonical form are introduced, and then decomposed to a product of unit upper and lower triangular or block-triangular matrices which define ladder structures. A fundamental approach to obtaining suitable unimodular matrices for one and two dimensions is to focus on the shift (translation) operators, as is done in the harmonic analysis discipline. Several matrix shift operators of different dimensions are introduced and their properties and applications are presented, most notable of which is that the McClellan transformation can be effected by a simple substitution of a 2×2 circulant matrix for the polynomial variable, $w = (z + z^{-1})/2$. Unimodular matrix groups and pertinent subgroups are identified, and these are observed to be subgroups of the special linear group over polynomials, $SL(k[w])$. A class of coiflet-like wavelets containing the well-known wavelet, based on the Burt and Adelson filter, is decomposed by these methods and is seen to require only 3/2 multiplications/sample if a scaling property, introduced herein, is satisfied. Making use of certain paraunitary wavelets, coiflets, that are closely comparable to the zero-phase wavelets of this class, it is seen that, in these cases, the zero-phase ladder algorithm is twice as fast as the paraunitary lattice algorithm.

Key Words: coiflet-like, zero phase, polyphase, quincunx, block-triangular, unimodular, group, ring, circulant

1. Introduction

Perfect reconstruction (PR) digital filter banks (DFB's) with unequal length, zero-phase filters are important in the realization of DFB's and wavelets [1–4], particularly for image processing, because of the desirability of linear-phase filtering in avoiding edge distortion and because the McClellan transformation [5, 6] can be used to obtain 2-dimensional filters. Ladder structures with two branches were recently introduced [7–11]. Design procedures for multi-branch ladders subsequently appeared with discussions of their application to implementing biorthogonal zero-phase DFB's and to image processing [12–16, 37]. The use of the McClellan transformation to obtain 2-dimensional DFB's of ladder structure was first discussed in [10, 11], and applications and extensions appeared in the subsequent references. The viewpoint that the ladder structure can be determined by decomposing suitable polynomial matrices will be emphasized, aspects of which were described in [13, 16]. The group structure of the polynomial matrices will be considered, and they will be seen to be subgroups of the special linear group of unimodular polynomial matrices [17]. Further discussions of the most relevant of these references will appear below.

---

* This paper is based upon the author's 1993 ICASSP paper, "U-L Block Triangular Matrix and Ladder Realizations of Subband Coders" [13].

Decompositions of para-unitary scattering matrices describing lossless analog networks [18], and related decompositions introduced by Vaidyanathan and his coworkers for paraunitary DFB's [19], and those introduced by Vetterli and LeGall for DFB's having linear-phase (symmetric) $H_0(z)$ and antisymmetric $H_1(z)$ of equal length, as in the orthogonal case [20], are well-known. A decomposition was also proposed in the latter reference for the class considered in this paper having both filters zero-phase or linear-phase, but it led to the conclusion that this class was less economical in terms of computational requirements than the other two classes.

A contribution of this paper is the demonstration that the ladder and block-triangular matrix decomposition makes the realization of the desirable class of DFB's with zero-phase filters more economical than either of the other classes, which, of course, is of particular importance for image processing. This result depends upon a scaling property [12, 14] to be discussed. A further contribution of this paper is the introduction of unimodular polyphase matrices describing one- and two-dimensional filter banks with zero-phase filters, with emphasis upon their properties as operators, the vectors upon which they operate, and their decomposition. It is seen that the polyphase matrix decomposition to Smith canonical form (SCF) yields a ladder and is analogous to the decompositions to lattice structures, but the factors are sparse, elementary triangular matrices, each requiring only one multiplication [13], as opposed to two for the lattice. A novel construction of these multidimensional polyphase matrices focuses on shift (or translation) operators, as done in the field of harmonic analysis [21], to obtain shift matrices of the correct dimension for the number of channels and the dimensionality [22, 8]. A result of this is that the McClellan transform is effected by a simple substitution of a matrix for a polynomial variable. A novel comparison of ladder and lattice structures using coiflet (paraunitary) and coiflet-like (zero-phase) designs introduced by Daubechies and her collaborators for image processing [4] is made which leads to the conclusion that the ladder requires only $1/2$ the number of computations of a comparable lattice. A compact, 3 step procedure, using the Euclidean algorithm in each step, for the design and ladder decomposition of these zero-phase DFB's, is described and illustrated for a DFB in which the low channel is a Burt and Adelson filter [23, 24].

The organization of the paper is as follows: One-dimensional shift operators and unimodular polyphase matrices are introduced in Section 2. In Section 3 the design and decomposition by the Euclidean algorithm is presented. In Section 4 matrix descriptions of DFB's for two-dimensional signals and their decomposition into block-triangular matrices, and corresponding ladders will be developed. In Section 5 higher dimensional shift operator matrices are briefly discussed, symmetrical ladders are introduced, and the computational efficiency of the ladder decomposition is examined in detail and compared with the lattice decomposition. Concluding comments follow in the summary.

2. Zero-Phase DFB's and Their Unimodular Matrix Representations

2.1. Change of Variable

It is well-known that zero-phase functions can be expressed as polynomials in a variable, say $w = (z + z^{-1})/2$ for the 1-D case, which becomes $w = \cos(\omega)$ for $z = e^{j\omega}$. Herrmann

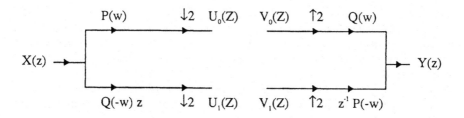

Figure 1. Analysis and synthesis PR digital filter banks assuming zero-phase filters.

introduced this variable change and a further change, $x = (1 - w)/2$, to design a class of maximally-flat FIR filters [25, 26]. In [4] the filters in this class are referred to as splines and designed with similar variable changes. In the variable w a PR two-channel DFB can be described as shown in Figure 1. For convenience, definitions $P(w) = H_0(z)$ and $Q(w) = G_0(z)$ have been made. Let $P(w)Q(w) = R(w)$. Then, for perfect reconstruction, $R(w)$ is a halfband function [27], and

$$R(w) + R(-w) = 2, \quad P(-w) = G_1(z), \quad \text{and} \quad Q(-w) = H_1(z), \tag{1}$$

as seen in Fig. 1. The explicit presence of z and z^{-1} simplifies the last two equations in (1) from which it can be seen that the identity DFB is given by $P(\pm w) = Q(\pm w) = 1$.

EXAMPLE 1 The lowest order DFB of this form with nontrivial $P(w)$ has $P(w) = 1 + w$ and $Q(w) = 1$. The regularity can be increased by choosing a new synthesis filter

$$Q'(w) = 1 + (w/2)(1 - w) = (1 + w)(1 - w/2), \tag{2}$$

which, like $P(w)$, has a zero at $w = -1$ corresponding to a double zero at $z = -1$. The halfband $R(w) = P(w)Q'(w)$ is maximally flat, but the two filters, factors of the halfband $R(w)$, do not have good, similar responses, a well-known problem. The design procedure in Section 3 provides a better way to design DFB's in which the lowpass filters $P(w)$ and $Q(w)$ have good regularity and approximately the same frequency responses.

2.2. *Matrix Representations and Properties of One-Dimensional DFB's*

In order that the shift-varying operations of up- and down-sampling in a DFB can be described, a single z-transform is insufficient, but the representations,

$$X(z) = X_0(z^2) + z^{-1}X_1(z^2) = X_0 + z^{-1}X_1, \tag{3}$$

of signals, in which the arguments, (z^2), are omitted for brevity in the last member and below, suggests several, equivalent, suitable representations [22]:

$$
\begin{aligned}
\mathbf{x} &= [X_0 \ X_1], \\
\mathbf{x}_c &= [X_0 \ z^{-1}X_1] = \mathbf{x}\mathbf{Z}^{-1}, \\
\mathbf{x}_f &= [X(z) \ X(-z)] = \mathbf{x}_c\mathbf{F}^*,
\end{aligned}
\tag{4}
$$

in which $\mathbf{Z} = \text{diag}(1, z)$ and $\mathbf{F}$ is the DFT matrix with rows, $\mathbf{f}_0 = [1\ 1]$ and $\mathbf{f}_1 = [1\ -1]$. For dimensions greater than 2, the conjugate transpose of $\mathbf{F}$, denoted by $*$, is appropriate for relating $\mathbf{x}_f$ to $\mathbf{x}_c$ as shown. The use of row vectors ensures that polyphase matrix factors, to be determined, will occur in the order, input on left and output on right, as in diagrams, rather than reversed order.

A simple, unifying viewpoint for matrix representations of DFB's is obtained by the observation that the basic operation of convolution is a linear combination of shift operations. Hence, corresponding to the above representations are shift operators corresponding to the variable, z. These operators are all based upon the circular shift matrix, $\mathbf{S} = \text{cdiag}(1, 1)$, a counter diagonal matrix for $M = 2$ (for general M, cdiag is replaced by a circulant matrix [28] for $\mathbf{T}_c$ or a related matrix $\mathbf{T}$ with an entry z^M since then $\mathbf{Z} = \text{diag}(1, z, \ldots, z^{M-1})$ is the generalization of $\mathbf{Z} = \text{diag}(1, z)$ used below [22]):

$$\begin{aligned}
\mathbf{T}_c &= z\mathbf{S} = \text{cdiag}(z, z), \\
\mathbf{T} &= \mathbf{Z}^{-1}\mathbf{T}_c\mathbf{Z} = \text{cdiag}(1, z^2), \\
\mathbf{T}_f &= (1/N)\mathbf{F}\mathbf{T}_c\mathbf{F}^* = \text{diag}(z, -z),
\end{aligned} \tag{5}$$

in which the entries in cdiag are in column order, i.e., from lower left to upper right, which is significant in the definition of $\mathbf{T}$. It is easily verified that these matrix operators represent z in $X(z)z$ for the foregoing representations of $X(z)$ in (4).

The representations $\mathbf{x}_c$ and $\mathbf{T}_c$ are the most convenient ones for use with zero-phase filters and will be used in this paper. (The subscript "c" originally meant clockwise [22], but in this paper it is convenient to attribute it to circulant.) The fact that $\mathbf{T}_c$ is a matrix that represents z in the expected manner is easily verified by observing that if $Y(z) = X(z)z$, then $Y(z) = X_1(z^2) + zX_0(z^2)$ implies $\mathbf{y}_c = [X_1\ zX_0]$ by the definition in (4). However, it is seen that $\mathbf{x}_c\mathbf{T}_c = \mathbf{y}_c$, so $\mathbf{T}_c$ has the same effect upon the vector $\mathbf{x}_c$ that z has upon $X(z)$. The two descriptions are reversible, hence isomorphic.

The matrix $\mathbf{T}_c$ is a circulant matrix and is diagonalized by the 2×2 DFT matrix, as shown in (5), giving $\mathbf{T}_f$. It is also a function of z so any z-transform, $H(z)$, is represented by $H(\mathbf{T}_c)$, including recursive functions. Since $H(z) = H_0(z^2) + z^{-1}H_1(z^2)$,

$$H(\mathbf{T}_c) = H_0(\mathbf{T}_c^2) + \mathbf{T}_c^{-1}H_1(\mathbf{T}_c^2) = \begin{bmatrix} H_0(z^2) & z^{-1}H_1(z^2) \\ z^{-1}H_1(z^2) & H_0(z^2) \end{bmatrix}, \tag{6}$$

a circulant matrix with z-transform entries. It can be seen when $H(z) = z$, that (6) gives $\mathbf{T}_c$ as previously defined.

From $\mathbf{T}_c$, a representation of $w = (z + z^{-1})/2$, $\mathbf{T}_w$, can be seen to be

$$\mathbf{T}_w = w\mathbf{S} = \text{cdiag}(w, w). \tag{7}$$

Let the polynomial $P(w) = P_0(w^2) + wP_1(w^2)$. Then the matrix polynomial, $P(\mathbf{T}_w)$,

$$P(\mathbf{T}_w) = \begin{bmatrix} P_0(w^2) & wP_1(w^2) \\ wP_1(w^2) & P_0(w^2) \end{bmatrix}, \tag{8}$$

is a circulant matrix. We wish to establish the following representation,

$$X(z)P(w) \rightarrow \mathbf{x}_c P(\mathbf{T}_w), \tag{9}$$

showing that the particular vector representation $\mathbf{x}_c$ is compatible with the zero-phase operator, $P(\mathbf{T}_w)$.

PROPERTY 1 The commutative matrix-polynomial ring, $\{P(\mathbf{T}_w)\}$, is a faithful (i.e., isomorphic [29]) representation of the polynomial ring $\{P(w)\}$.

Proof: Commutativity and the ring structure of matrix polynomials is well-known [30], so the main thing to prove is that the scalar- to matrix-polynomial mapping, $\mathcal{M}$, is an isomorphism. By virtue of (8), $P(w) = [1 \ 0]P(\mathbf{T}_w)[1 \ 1]^T$, so the kernel of $\mathcal{M}$, the set of $P(w)$ for which $P(\mathbf{T}_w)$ is the zero matrix, consists only of $P(w) = 0$, establishing that the mapping is an isomorphism. (Note that the degree of $P(\mathbf{T}_w)$ is that of $P(w)$, and that division, the Euclidean algorithm, cancelation of factors, etc. can all be done in $\{P(\mathbf{T}_w)\}$).

The following property can be easily verified:

PROPERTY 2 $P(\mathbf{T}_w)$ is a circulant matrix with polynomial entries which represents $P(w)$ in $X(z)P(w)$ if $X(z)$ is represented by $\mathbf{x}_c$.

An interesting consequence of this is that $P(\mathbf{T}_w)$ has important circulant matrix properties such as being diagonalized by the DFT matrix, $\mathbf{F}$. However, it represents linear, not circular, convolution by virtue of its polynomial entries. With entries from a field, a Toeplitz or a high order circulant matrix approximation would be required to describe linear convolution.

Let $\mathbf{v}_c$ and $\mathbf{y}_c$ represent the input and output signals of a synthesis DFB and define the synthesis DFB polyphase matrix, $\mathbf{G}_c$, by $\mathbf{y}_c = \mathbf{v}_c \mathbf{G}_c$. Note that $Y(z)$ is described by $\mathbf{y}_c$. From Fig. 1 it is seen that $G_0(z) = Q(w)$ and $G_1(z) = P(-w)$, giving $\mathbf{G}_c$ by letting its first row, $\mathbf{G}_{c,o}$, be the first row of $Q(\mathbf{T}_w)$ and the second row, $\mathbf{G}_{c,1}$, be the second row of $P(-\mathbf{T}_w)$, the second row being chosen because of the z^{-1} which presents the operator $P(-\mathbf{T}_w)$ with input vector $[0 \ z^{-1}V_1(z^2)]$. Thus, the polyphase synthesis DFB matrix is

$$\mathbf{G}_c = \begin{bmatrix} \mathbf{G}_{c,0} \\ \mathbf{G}_{c,1} \end{bmatrix} = \begin{bmatrix} Q_0(w^2) & wQ_1(w^2) \\ -wP_1(w^2) & P_0(w^2) \end{bmatrix}, \tag{10a}$$

and, by interchanging $P(w)$ and $Q(w)$, $\mathbf{H}_c$ is

$$\mathbf{H}_c = \begin{bmatrix} \mathbf{H}_{c,0} \\ \mathbf{H}_{c,1} \end{bmatrix} = \begin{bmatrix} P_0(w^2) & wP_1(w^2) \\ -wQ_1(w^2) & Q_0(w^2) \end{bmatrix}, \tag{10b}$$

The analysis DFB is described by $\mathbf{H}_c^T$, the superscript ^{T} denoting transpose, if the signal inputs are row vectors, as assumed here; alternatively, if the signal inputs are column vectors, the same two matrices describe them, but then $\mathbf{G}_c^T$ describes the synthesis DFB.

The pair of DFB's and their signals described by $\mathbf{x}_c \mathbf{H}_c^T = \mathbf{u}_c$ and $\mathbf{v}_c \mathbf{G}_c = \mathbf{y}_c$ suggest that an alternative to Figure 1, Figure 2, describes the DFB pair, and it makes evident that the

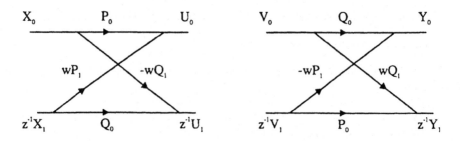

Figure 2. Analysis and Synthesis DFB's with Zero-Phase Filters. The signals are $X_i = X_i(z^2)$, etc. and the DFB functions are $P_i = P_i(w^2)$ and $Q_i = Q_i(w^2)$.

analysis and synthesis DFB's have identical structure. It will be convenient to illustrate the ladder structures to be developed with similar diagrams. From (1) it can be seen that

$$P(w)Q(w) + P(-w)Q(-w) = 2. \tag{11}$$

Some properties of the above matrices follow:

PROPERTY 3   The matrix $\mathbf{G}_c$ defined from a pair $\{P(w), Q(w)\}$ which satisfies (11) is a unimodular polynomial matrix. The inverse matrix is the transpose of the matrix, $\mathbf{H}_c$, formed in the same manner as $\mathbf{G}_c$ with $P(w)$ and $Q(w)$ interchanged, i.e.,

$$\mathbf{G}_c^{-1} = \mathbf{H}_c^{\mathrm{T}}. \tag{12}$$

The set of all such $\mathbf{G}_c$ together with the operation of matrix multiplication is a group.

Proof: The first statement is true if $|\mathbf{G}_c| = Q_0(w^2)P_0(w^2) + w^2 Q_1(w^2)P_1(w^2) = 1$; this follows directly from (11). The second statement is verified by showing $\mathbf{H}_c^{\mathrm{T}}\mathbf{G}_c = \mathbf{I}$. The last statement is true if each inverse matrix is of the same form as the matrix itself. Comparing $\mathbf{H}_c^{\mathrm{T}}$ with $\mathbf{G}_c$, it is seen that by defining $A(w) = P_0(w^2) + (-w Q_1(w^2))$ and $B(w) = Q_0(w^2) + (-w P_1(w^2))$ and using the pair $\{A(w), B(w)\}$ to construct a matrix of the form $\mathbf{G}_c$, that (12) has the same form as $\mathbf{G}_c$. Therefore, the set of all such $\mathbf{G}_c$ contains the inverses for each member of the set, the identity $\mathbf{I}$ is in this set since it has the required form, and the group operation is associative since it is matrix multiplication. All the group axioms are satisfied showing that the last statement is true.

A corollary is that analysis and synthesis DFB's are described by exactly the same form of unimodular matrix and are therefore both in this group. It will be convenient to name the group.

Definition. A 2×2 polynomial matrix has the zero-phase unimodular (ZPU) property if the diagonal entries are polynomials in w^2 including the zero power, the off diagonal entries are polynomials in odd powers of w, and the matrix is unimodular. The set of all such matrices is the ZPU group, a subgroup of the group of all unimodular polynomial matrices, $SL(k[w])$, where k is the coefficient field.

A scaling, to be discussed in the next section, of a pair $\{P(w), Q(w)\}$ defining a $\mathbf{G}_c$ is obtained with $P(0) = 1 = Q(0)$. This implies that the resulting ZPU matrix satisfies $\mathbf{G}_c(w)|_{w=0} = \mathbf{I}$.

PROPERTY 4 The set of all ZPU matrices with the following restriction, $\{\mathbf{G}_c(w)|\mathbf{G}_c(0) = \mathbf{I}\}$, is a subgroup of the group of ZPU matrices.

Proof: Associativity is inherited from the group, the identity $\mathbf{I}$ and the inverses are in the subset, and the subset is closed under matrix multiplication establishing the property.

Denote this subgroup by ZPU1. Since the restriction saves multiplications, all designs considered here, including those already mentioned, are in ZPU1. The following property shows that the reduction to SCF can be done with ZPU1 operations.

PROPERTY 5 Any matrix in ZPU1 can be reduced to its Smith canonical form, $\mathbf{I}$, by premultiplying by elementary matrices in ZPU1 which are alternately unit upper and lower triangular.

Proof: Postmultiplying a ZPU1 matrix, say $\mathbf{M}$, by $[1\ 1]^T$ gives a column vector of the form $[A(w)\ B(w)]^T$. Premultiplying by alternating unit upper and lower triangular, elementary ZPU1 matrices performs the Euclidean algorithm on the pair, ultimately reducing one to a constant, since the pair of polynomial matrices defining any ZPU1 matrix can be seen to be relatively prime (since its determinant $= 1$). Applying these elementary matrices to $\mathbf{M}$ therefore reduces it to a unit triangular matrix, and applying its inverse is an added step that results in the unit matrix, which is therefore the SCF of $\mathbf{M}$.

EXAMPLE 2 The polynomials $P(w)$ and $Q'(w)$ of Example 1 define $\mathbf{G}_c$ and its reduction,

$$\mathbf{H}_c^T \mathbf{G}_c = \begin{bmatrix} 1 & 0 \\ w & 1 \end{bmatrix} \begin{bmatrix} 1 & -w/2 \\ 0 & 1 \end{bmatrix} \begin{bmatrix} 1 - w^2/2 & w/2 \\ -w & 1 \end{bmatrix} = \mathbf{I} \tag{13}$$

gives $\mathbf{H}_c^T = \mathbf{G}_c^{-1}$ in factored form. From this it is seen that

$$\mathbf{G}_c = \begin{bmatrix} 1 & w/2 \\ 0 & 1 \end{bmatrix} \begin{bmatrix} 1 & 0 \\ -w & 1 \end{bmatrix}, \tag{14}$$

and the factors are in the expected order for operating on signal row vectors.

3. The Design of DFB's with High Regularity and Closely Matching Zero-Phase Filters Using the Euclidean Algorithm

3.1. Design With The Euclidean Algorithm

Prior to the ladder expansion a desirable design must be available. The design is determined by a pair of zero-phase polynomials, $\{P(w), Q(w)\}$, with product, $R(w)$, which is halfband

and for which each is a desirable lowpass filter, i.e.,

$$P(w) \cong \sqrt{R(w)} \cong Q(w). \tag{15}$$

The procedure to be described is that of [4], which provides the desired regularity, with an added scaling restraint on the product, $P(w)Q(w)$, namely $P(0) = Q(0)$ which saves multiplications and is consistent with (15). Because of the smoothness resulting from the regularity it suffices to specify

$$P(0) = Q(0) = 1 \quad \text{and} \quad P(1) = Q(1) = \sqrt{2}. \tag{16}$$

The Euclidean algorithm [31] will be used in both of the following steps. Its use in DFB design is well-known [2, 32] and was mentioned as an alternative approach in [4].

STEP 1 A maximally flat, halfband filter, $R'(w)$, is obtained by solving

$$2 = R'(w) + R'(-w) = (1 + w)^k X(w) + (1 - w)^k X(-w), \tag{17}$$

for the unique unknowns $X(w)$ and $X(-w)$. The results for $k = 1 - 4$ are

$$X(w) = 1, \, 1 - w/2, \, 1 - (9/8)w + (3/8)w^2, \, 1 - (29/16)w + (5/4)w^2 - (5/16)w^3, \tag{18}$$

agreeing with [26]. Non-unique higher degree solutions will be used in the next step.

EXAMPLE 3: DESIGN OF A MAXIMALLY FLAT FILTER Assuming $k = 2$, let $A_0(w) = (1 + w)^2$ and $A_1(w) = (1 - w)^2$. Then 2 steps of the Euclidean algorithm suffice to yield a constant remainder and are described by the two triangular matrices in the right member of

$$\begin{bmatrix} A_2(w) \\ A_3(w) \end{bmatrix} = \begin{bmatrix} 1 & 0 \\ -\left(\dfrac{w}{4} - \dfrac{1}{2}\right) & 1 \end{bmatrix} \begin{bmatrix} 1 & -1 \\ 0 & 1 \end{bmatrix} \begin{bmatrix} A_0(w) \\ A_1(w) \end{bmatrix}, \tag{19}$$

in which $A_2(w) = 4w$ and $A_3(w) = 1$. Then $A_3(w)$ is seen to be

$$1 = A_0(w) \left(\frac{1}{2} - \frac{w}{4}\right) + A_1(w) \left(\frac{1}{2} + \frac{w}{4}\right). \tag{20}$$

Scaling up by a factor of 2 in accordance with (17) gives the $X(w)$ in (18).

STEP 2 In order to satisfy the conditions (16) and retain flatness at origin choose

$$P(w) = (1 + w)^k (X(w) + (\sqrt{2} - 1)(1 - w)^k)/\sqrt{2}, \quad \text{and} \quad Q(w) = (1 + w)^k Y(w), \tag{21}$$

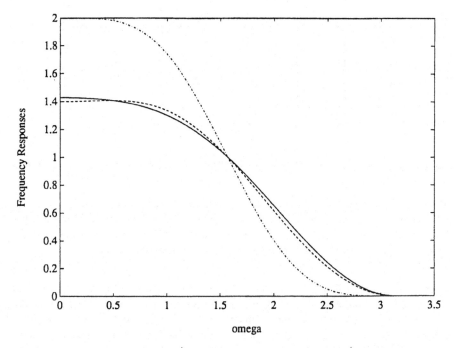

Figure 3. The Burt and Adelson filter, $H_0(e^{j\omega}) = P(w)$ ———, its complement, $G_0(e^{j\omega}) = Q(w)$ - - - - - -, and their halfband product filter, $H_0(e^{j\omega})G_0(e^{j\omega}) = N_0(e^{j\omega}) = R(w)$.

in which the unknown $Y(w)$ is found to make $Q(w)$ complementary to $P(w)$ by solving

$$2 = P(w)(1 + w)^k Y(w) + P(-w)(1 - w)^k Y(-w). \tag{22}$$

The following example illustrates this design algorithm.

EXAMPLE 4: A DFB INCORPORATING A BURT AND ADELSON FILTER It was observed in [3, 4] that the filter class introduced by Burt and Adelson [23, 24] has the z-transform

$$H(z) = (.25 - .5a)z^2 + .25z + a + .25z^{-1} + (.25 - .5a)z^{-2}, \tag{23}$$

which can be scaled to have the desired form for $k = 1$,

$$P(w) = (1 + w)(1 - bw) = 1 + (1 - b)w - bw^2, \tag{24}$$

with $b = 1 - 1/\sqrt{2}$. A rational approximation, $b = 2/7$, corresponds to the choice $a = .6$ in [3, 4]. Using this in (22) gives $Y(w) = (-.12w^2 - .18w + 1)$ and

$$Q(w) = -.12w^3 - .3w^2 + .82w + 1. \tag{25}$$

Figure 3 illustrates the frequency responses of these filters and their product, $R(w)$, a

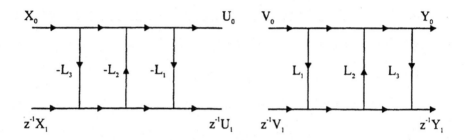

Figure 4. Ladder Realization of the Analysis and Synthesis DFB's in which $P(w)$ is a Burt and Adelson Filter.

halfband filter. Since the halfband filter has all even powers of z and w equal to zero, except the zero power, its Fourier series consists of a constant and odd powers of $\cos(\omega)$ and has odd symmetry about $\omega = \pi/2$, as shown. The relation (15) is seen to be well satisfied, the rational approximation of b resulting in only a small discrepancy at $\omega = 0$.

EXAMPLE 5: LADDER REALIZATION OF THE DFB WITH A BURT AND ADELSON FILTER
The DFB of Example 4 will be used to illustrate the matrix decomposition and ladder structure. Execute the Euclidean algorithm upon $Q(w) = A_0(w)$ and $P(-w) = A_1(w)$ in the example with the added step, mentioned in the proof of Property 5, to get the unknown quotients, denoted $L_i(w)$, which are the ladder branch coefficients of the synthesis filter:

$$
\begin{aligned}
A_0(w) &= L_1(w)A_1(w) + A_2(w) = (.42w)A_1(w) + (.4w + 1) \\
A_1(w) &= L_2(w)A_2(w) + A_3(w) = ((-5/7)w)A_2(w) + 1 \\
A_2(w) &= L_3(w)A_3(w) + A_4(w) = .4w + 1.
\end{aligned}
\tag{26}
$$

The Euclidean algorithm terminates with the second equation since the remainder is $A_3(w) = 1$, and the third equation is the extra step from which the last coefficient is obtained. The quotients are off diagonal elements of ZPU1 matrices and are necessarily odd. The resulting ladder realization of the analysis and synthesis DFB's is given in Figure 4. For additional higher order ladder examples satisfying (15) see [12, 37], and for applications to image compression, see [14]. In the two higher order cases shown in [12, 37] similar responses to those in Figure 3 with significantly sharper cutoffs are seen. The quotients, $L_i(w)$, for the higher order cases illustrate more general cases of the Euclidean algorithm, in that in each case there is one quotient polynomial which is of higher odd degree than one, since in these cases deg $Q(w)$ exceeds deg $P(w)$ by more than one.

4. Two-Dimensional Polyphase Matrices

4.1. Two-Dimensional Variables

Define vertical and horizontal variables, $w_v = (z_v + z_v^{-1})/4$ and $w_h = (z_h + z_h^{-1})/4$. The variable change, $w = w_v + w_h$ can be made to obtain 2-dimensional zero-phase filters, which corresponds to the use of the McClellan transformation. This particular 2-dimensional transformation is useful for obtaining diamond support filters for use in decimating to quincunx lattices [1, 6] and is a frequently used example [10, 11]. The resulting diamond-shaped point-spread matrix of $w = ((z_v + z_v^{-1} + z_h + z_h^{-1})/4)$ is

$$\mathbf{D} = 1/4 \begin{bmatrix} 0 & 1 & 0 \\ 1 & 0 & 1 \\ 0 & 1 & 0 \end{bmatrix}. \tag{27}$$

EXAMPLE 6: FREQUENCY TO SPATIAL DOMAIN From Example 1 (dropping the primes) it is seen that the resulting halfband function is, $R(w) = (1 + w)^2(1 - w/2)$. Assuming $w = (w_v + w_h)$, $R(w)$ can be converted to the spatial domain by replacing w with $\mathbf{D}$ in (27) and replacing polynomial multiplication of w by 2-dimensional convolution resulting in

$$\mathbf{R} = (\mathbf{I} + \mathbf{D}) * (\mathbf{I} + \mathbf{D}) * (\mathbf{I} - (1/2\mathbf{D})) \tag{28}$$

where $*$ indicates 2-dimensional convolution. Equation (28) can be seen to be the halfband point-spread matrix with diamond support [10],

$$\mathbf{R} = \frac{1}{128} \begin{bmatrix} 0 & 0 & 0 & -1 & 0 & 0 & 0 \\ 0 & 0 & -3 & 0 & -3 & 0 & 0 \\ 0 & -3 & 0 & 39 & 0 & -3 & 0 \\ -1 & 0 & 39 & 128 & 39 & 0 & -1 \\ 0 & -3 & 0 & 39 & 0 & -3 & 0 \\ 0 & 0 & -3 & 0 & -3 & 0 & 0 \\ 0 & 0 & 0 & -1 & 0 & 0 & 0 \end{bmatrix}. \tag{29}$$

The presence of diamonds of zeros that alternate with nonzero diamonds indicates that this 2-dimensional impulse response has the halfband property. Its factors are shown in (28) and would be difficult to determine or non-existent were it not for its underlying 1-dimensional origin.

4.2. Representations and Properties of Two-Dimensional DFB's

The results of Section 2.2 may be extended to the 2-D case. The result is a surprizingly simple generalization of the 1-dimensional case obtained by replacing the variable, w, with a matrix, $\mathbf{W}$. The McClellan transformation will be seen to be implemented by this simple substitution in the shift operator of (7) and therefore in the resulting 2-dimensional polyphase matrices. Consider representing 2-dimensional signals in terms of 4 rectangular cosets as follows:

$$X(z_v, z_h) = \{X_{00}(z_v^2, z_h^2) + z_v^{-1}z_h^{-1}X_{11}(z_v^2, z_h^2)\} + \{z_h^{-1}X_{01}(z_v^2, z_h^2) + z_v^{-1}X_{10}(z_v^2, z_h^2)\}. \tag{30}$$

The four $X_{ij} = X_{ij}(z_v^2, z_h^2)$ are 2-D signals on rectangular cosets. The pair of cosets in curly brackets describes signals on two quincunx cosets [1, 6]. This representation of quincunx cosets with rectangular subcosets is an alternative to those of Viscito and Allebach and Tay and Kingbury [33, 34] leading to the properties below: Define

$$\mathbf{x}_c = [X_{00} \ z_v^{-1}z_h^{-1}X_{11} \ z_h^{-1}X_{01} \ z_v^{-1}X_{10}],$$
$$\mathbf{x} = \mathbf{x}_c \mathbf{Z} = [X_{00} \ X_{11} \ X_{01} \ X_{10}], \tag{31}$$

length 4 vectors with $\mathbf{Z} = \text{diag}(1, z_v z_h, z_h, z_v)$. These representations can be seen to be 2-dimensional generalizations of (4), and the same names are used for convenience. The ordering of the polyphase components of $X(z_v, z_h)$ places the two quincunx cosets in evidence and gives the following block-circulant matrix generalization of (7),

$$T_{\mathbf{W}} = \text{cdiag}(\mathbf{W}, \mathbf{W}) = \begin{bmatrix} \mathbf{0} & \mathbf{W} \\ \mathbf{W} & \mathbf{0} \end{bmatrix}, \text{ and} \tag{32a}$$

$$\mathbf{W} = \text{diag}(w_h, w_h) + \text{cdiag}(w_v, w_v) = \begin{bmatrix} w_h & w_v \\ w_v & w_h \end{bmatrix}. \tag{32b}$$

Letting $P(\mathbf{W}) = P_0(\mathbf{W}^2) + \mathbf{W}P_1(\mathbf{W}^2)$,

$$P(T_{\mathbf{W}}) = \begin{bmatrix} P_0(\mathbf{W}^2) & \mathbf{W}P_1(\mathbf{W}^2) \\ \mathbf{W}P_1(\mathbf{W}^2) & P_0(\mathbf{W}^2) \end{bmatrix}. \tag{33}$$

The following representation is to be shown:

$$X(z_v, z_h)P(w) \rightarrow \mathbf{x}_c P(T_{\mathbf{W}}). \tag{34}$$

Since $\mathbf{W}$ is a circulant matrix, any polynomial in $\mathbf{W}$ is a circulant matrix. A matrix with blocks which are circulant matrices will be denoted a circulant-block matrix, and one with blocks which have the circulant ordering, i.e., each row of blocks after the first is a right circular shift of the preceding row, will be denoted a block-circulant matrix following Davis [28]. Matrices with both properties are referred to as doubly-block-circulant matrices [35]. The following property results:

PROPERTY 6 The commutative doubly-block-circulant matrix-polynomial ring, $\{P(T_{\mathbf{W}})\}$, is a faithful representation of the polynomial ring $\{P(w)\}$.

This representation with 4×4 matrices is another representation, like the representation of Property 1 with 2×2 matrices, of the same polynomial ring, the distinction being that the matrix dimension and structure, as determined by the shift operator matrix, T_w or $T_{\mathbf{W}}$, is appropriate for the sampling lattice upon which the signals are defined. Observing that $P(T_{\mathbf{W}})[1 \ 1 \ 1 \ 1]^T = P(w)[1 \ 1 \ 1 \ 1]^T$, it is seen that the property can be established as before and the degree of $P(T_{\mathbf{W}})$ is that of $P(w)$, etc. The following property is established by verification of (34).

PROPERTY 7 $P(T_W)$ is a doubly-block-circulant matrix with polynomial blocks, i.e., they are polynomials in the circulant matrix $\mathbf{W}$. $P(\mathbf{T_W})$ represents $P(w)$ in $X(z_v, z_h)P(w)$ if $X(z_v, z_h)$ is represented by $\mathbf{x}_c$ in (31).

Let $\mathbf{u}_c$ and $\mathbf{y}_c$ represent the input and output signals of a 2-dimensional synthesis DFB and define the synthesis DFB matrix, $\mathbf{G}_c$, by $\mathbf{y}_c = \mathbf{u}_c \mathbf{G}_c$. Note that $Y(z_v, z_h)$ is described by $\mathbf{y}_c$. $\mathbf{G}_c$ is obtained from $G_0(z) = Q(w)$ and $G_1(z) = P(-w)$, by letting its first block row, $\mathbf{G}_{c,o}$, be the first block row of $Q(\mathbf{T_W})$ and the second block row, $\mathbf{G}_{c,1}$ be the second block row of $P(-\mathbf{T_W})$, with $\mathbf{T_W}$ given in (32). Thus,

$$\mathbf{G}_c = \begin{bmatrix} \mathbf{G}_{c,0} \\ \mathbf{G}_{c,1} \end{bmatrix} = \begin{bmatrix} Q_0(\mathbf{W}^2) & \mathbf{W}Q_1(\mathbf{W}^2) \\ -\mathbf{W}P_1(\mathbf{W}^2) & P_0(\mathbf{W}^2) \end{bmatrix}, \tag{35a}$$

and, by interchanging $P(\mathbf{W})$ and $Q(\mathbf{W})$, $\mathbf{H}_c$ becomes

$$\mathbf{H}_c = \begin{bmatrix} \mathbf{H}_{c,0} \\ \mathbf{H}_{c,1} \end{bmatrix} = \begin{bmatrix} P_0(\mathbf{W}^2) & \mathbf{W}P_1(\mathbf{W}^2) \\ -\mathbf{W}Q_1(\mathbf{W}^2) & Q_0(\mathbf{W}^2) \end{bmatrix}. \tag{35b}$$

It is seen that $\mathbf{G}_c$ and $\mathbf{H}_c$ are special circulant-block matrices since the blocks are polynomials in a circulant matrix. These and their block-triangular transforming matrices are a generalization of ZPU1, ZPU1B, obtained by substituting $\mathbf{W}$ for w. It is seen that the 1-dimensional DFB description has effectively been extended to 2 dimensions by this substitution in the DFB matrices of (10). The conclusions of the preceding properties and discussions will be summarized in the following theorem, noting that the internal circulant-block matrices which are also polynomial matrices become polynomials in the 1-dimensional case:

THEOREM Perfect reconstruction analysis and synthesis 1- and 2-dimensional DFB's with zero-phase filters are described by unimodular, circulant-block, polyphase matrices, $\mathbf{H}_c^T$ and $\mathbf{G}_c$, defined in (35) which satisfy $\mathbf{H}_c^T \mathbf{G}_c = \mathbf{G}_c \mathbf{H}_c^T = \mathbf{I}$. These matrices are in ZPU1B, and they are reduced to their Smith canonical form, $\mathbf{I}$, by circulant-block, unit upper and lower block-triangular matrices. The resulting decompositions of $\mathbf{H}_c$ and $\mathbf{G}_c$ yield 1- and 2-dimensional ladder realizations of the DFB's.

Proof: The proof that $\mathbf{H}_c^T$ and $\mathbf{G}_c$ represent the DFB's may be obtained by verification of $\mathbf{u}_c = \mathbf{x}_c \mathbf{H}_c^T$ and $\mathbf{y}_c = \mathbf{u}_c \mathbf{G}_c$ for the 2-dimensional case and the remaining statements are obvious generalizations of their 1-dimensional counterparts.

The conversion of these results to z-transform notation can be conveniently done by letting $\mathbf{Z}$ be defined by $\text{diag}(1, z)$ and $\text{diag}(1, z_v z_h, z_h, z_v)$ for the 1- and 2-D cases, respectively. A superscript P, ^{P}, will denote the para-Hermitian transpose of a matrix obtained by taking the Hermitian transpose with the generalization that the variable z is replaced by z^{-1} rather than treating it as a complex number which would then be conjugated [18]. Defining $\mathbf{H}^P = \mathbf{Z}^{-1}\mathbf{H}_c^T\mathbf{Z}$ and $\mathbf{G} = \mathbf{Z}^{-1}\mathbf{G}_c\mathbf{Z}$, alternative forms of polyphase matrices with $\mathbf{H}^P\mathbf{G} = \mathbf{G}\mathbf{H}^P = \mathbf{I}$ are obtained. The transformation by $\mathbf{Z}$ and $\mathbf{Z}^{-1}$ results in unit upper and lower block-triangular matrix factors and ladder realizations of $\mathbf{H}$ and $\mathbf{G}$, which operate on vectors $\mathbf{x}$ in (9) and (39). This alternative representation with $\mathbf{H}^P$ and $\mathbf{G}$ is useful since it leads to a description of these results in terms of z-transforms, and the conversion to this

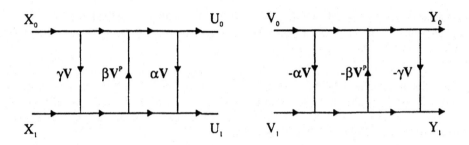

Figure 5. Analysis and Synthesis Ladder Realization for the DFB's Incorporating a Burt and Adelson Filter in which Signals and Branch Coefficients are 1- and 2-Dimensional Z-transforms.

form is a useful preliminary step to implementation. It is easily verified that transformation of the unit block-triangular matrices in $\mathbf{W}$ (or w) by $\mathbf{Z}$ still results in unit block-triangular matrices, but in the variables Z_v and Z_h (or Z). The following example identifies the new, unit block-triangular matrices:

EXAMPLE 7: 2-DIMENSIONAL LADDERS FOR THE DFB WITH A BURT AND ADELSON FILTER
From Figure 4, the matrix product describing the analysis filter, $\mathbf{H}_c^T$, and its transformation to $\mathbf{H}^P = \mathbf{Z}^{-1}\mathbf{H}_c^T\mathbf{Z}$ with $\mathbf{Z} = \mathrm{diag}(1, z_v z_h, z_h, z_v)$ (or $\mathrm{diag}(1, z)$) follow:

$$\mathbf{H}^P = \mathbf{Z}^{-1}\begin{bmatrix} \mathbf{I} & \gamma\mathbf{W} \\ 0 & \mathbf{I} \end{bmatrix}\begin{bmatrix} \mathbf{I} & 0 \\ \beta\mathbf{W} & \mathbf{I} \end{bmatrix}\begin{bmatrix} \mathbf{I} & \alpha\mathbf{W} \\ 0 & \mathbf{I} \end{bmatrix}\mathbf{Z} = \begin{bmatrix} \mathbf{I} & \gamma\mathbf{V} \\ 0 & \mathbf{I} \end{bmatrix}\begin{bmatrix} \mathbf{I} & 0 \\ \beta\mathbf{V}^P & \mathbf{I} \end{bmatrix}\begin{bmatrix} \mathbf{I} & \alpha\mathbf{V} \\ 0 & \mathbf{I} \end{bmatrix}, \quad (36)$$

in which $\alpha = -.42$, $\beta = 5/7$, and $\gamma = -.4$. It is seen that the convenient unit triangular property is preserved under the illustrated tranformation. In the 1-dimensional case,

$$\mathbf{V} = \frac{Z+1}{2} \text{ and } \mathbf{V}^P = \frac{1+Z^{-1}}{2}, \quad Z = z^2, \tag{37a}$$

and, in the 2-dimensional case,

$$\mathbf{V} = \frac{1}{4}\begin{bmatrix} Z_h+1 & Z_v+1 \\ 1+Z_v^{-1} & 1+Z_h^{-1} \end{bmatrix} \text{ and } \mathbf{V}^P = \frac{1}{4}\begin{bmatrix} 1+Z_h^{-1} & Z_v+1 \\ 1+Z_v^{-1} & Z_h+1 \end{bmatrix}, \quad Z_h = z_h^2, Z_v = z_v^2, \tag{37b}$$

first defined in [13]. The ladder in Figure 5 results. In the 2-dimensional case, $X_0 = [X_{00}\ X_{11}]$ is the signal on the quincunx sublattice, and $X_1 = [X_{01}\ X_{10}]$ is the signal on its complementary coset, with similar interpretations for the other signals. All signals and operators are functions of the low-rate variables, $Z = z^2$, or $Z_v = z_v^2$ and $Z_h = z_h^2$.

For implementation purposes it is helpful to note that individual blocks in each of the factors in (36) are ordinary z-transforms, either in w or z^2 for the one dimensional case or in $w = w_v + w_h$ or z_v^2 and z_h^2 in the 2-dimensional case, so they can be inverse z-transformed, as in (28), to very simple impulse responses in accordance with (37). If the signal or image is to be processed sequentially by blocks, it may be convenient to make the ladders causal or restricted to a quarter plane by introducing constant matrices, $Z^{-1}I_2$ or $Z_v^{-1}Z_h^{-1}I_4$, where needed, for the 1- and 2-dimensional cases respectively, in the matrix description, or with equivalent operations directly on the signal flow graph.

5. Discussion

Shift Operators and Representations: A unique and useful aspect of this paper is the emphasis upon the shift operators and their representations. Several examples for 1-D and 2-D cases were considered. The novel approach of representing the shift operator, and hence convolution, with a matrix of the appropriate dimension, 2 for the 1-D case and 4 for the 2-D case was discussed and illustrated. When more resolution levels or channels are to be considered, active areas of research, higher dimensional operators can be used. In [8] an 8×8 shift operator was used to illustrate 3 levels of resolution reduction, and a general method for DFB's with longer filter lengths was presented.

It has been seen that a careful choice of vectors for signal representation, including the use of four rectangular cosets to describe two quincunx cosets, has resulted in 1- and 2-dimensional matrices of the same structure which are related by the McClellan transform. The reduction of these matrices to canonical form can be done using the Euclidean algorithm on the 1-dimensional prototype, and 1- and 2-dimensional ladder structure realizations of the DFB's are obtained. Separable processing could also be used, so 2-dimensional DFB's with rectangular or diamond support filters can both be implemented, perhaps all that would be required in most image processing applications.

The reason for this convenient similarity between the 1- and 2-dimensional cases is that in both dimensions the signal up and down sampling is done on two complementary subcosets of the rectangular sampling lattice leading to a two channel DFB in both cases. Conveniently, the McClellan transformation can then be effected by simply substituting a 2×2 circulant matrix, $\mathbf{W}$, for the scalar variable, $w = (z + z^{-1})/2$. A generalization of this convenient result to other McClellan transforms and the generalization to more channels are interesting, related problems, not treated here, but which are under investigation.

Symmetrical Ladders: It may be seen that choosing $b = 1 - 1/\sqrt{2}$ in Example 4, instead of the rational approximation, gives ladder coefficients which are symmetrical with respect to the center of the sequence. It can be shown that such a symmetric ladder results if and only if $Q_0(w^2) = P_0(w^2)$ and that then $P_0^2(w^2) + w^2 P_1(w^2)Q_1(w^2) = 1$. Observing that $1 - P_0^2(w^2) = (1 + P_0(w^2))(1 - P_0(w^2)) = w^2 P_1(w^2)Q_1(w^2)$, an interesting constraint is placed in evidence. The existence of symmetrical ladders with good responses other than that given is an open question.

The Fast Algorithm: The number of parameters, and hence multipliers, is greatly reduced by the decomposition yielding a fast algorithm for subband coding. For comparison it can be observed that each of the three, biorthogonal, zero-phase wavelet designs presented in [4], one of which includes the Burt and Davidson filter, has an associated orthogonal wavelet design which is closely related in both order and response to the zero-phase design. The orthogonal filter designs are referred to as "coiflets" in the reference, and were specifically designed by Daubechies in response to a request from Coifman, for whom these designs were named, to closely approximate zero-phase filters [36]. They provide a convenient way to compare the computational efficiency of the ladders with the lattice decompositions of the other classes.

Let the length of the coiflets be $L = 2K$. For the case illustrated in Example 5, $K = 3$, and for the higher order cases, $K = 6$ and 9. In these cases the number of coefficients in the ladder branches of the biorthogonal wavelet with coiflet-like filters for the 1-dimensional case is also K [12, 37]. Since two input samples are processed by each DFB consisting of K coefficients, the number of multiplications is $K/2$ per sample. This favorable result requires that the scaling property, $P(0) = Q(0) = 1$ be satisfied [12]. It is shown by Vetterli and LeGall [20] that the number of multiplications per sample for a decomposed paraunitary DFB (coiflet) is $K + 1$. The same is true for the decomposed DFB with one symmetric (linear-phase) and one antimetric filter, except that multiplications can be traded for additions[1] in an alternative structure giving $(K + 1)/2$ multiplications per sample. Although this is comparable to the result in this paper, the significant increase in additions and the presence of an antimetric high channel filter is undesirable. Both the zero-phase and coiflet designs have closely comparable filter responses, so the $K/2$ multiplications/sample versus $K+1$, the exact zero-phase property, and the applicability of the McClellan transform appears to make the zero-phase designs advantageous in image processing.

Summary

The subgroup of perfect reconstruction DFB's based upon zero-phase filters, introduced herein, has been shown to admit ladder and block-triangular representations which are computationally efficient and of simple structure. The number of multiplications/sample is small for properly scaled filters, e.g., for the illustrated zero-phase wavelet containing a Burt and Adelson filter, it is only 3/2. The matrix representations of 2-dimensional convolution and of 2-dimensional DFB's and wavelets is conveniently obtained from the 1-dimensional representations by the scalar to matrix substitution, $w \rightarrow \mathbf{W}$, which implements the McClellan transformation from 1-dimensional filters to 2-dimensional filters with diamond support. The matrix representations were seen to be conveniently changed by matrix transformation which, as illustrated in Example 7 for the Burt and Adelson filter, allows the design which was made in the convenient representation with variable w and then $\mathbf{W}$ to be easily transformed to one more convenient for implementation.

Acknowledgment

The author wishes to acknowledge many helpful discussions with Drs. R. E. Van Dyck and S. J. Orfanidis and to express his appreciation to the reviewers for their detailed reviews. He greatly appreciates the encouragement of special issue editor, Dr. Sankar Basu, to write this paper.

Notes

1. Trades of several additions for one of the lattice multiplications are well known and increase computational load for most digital signal processors.

References

1. R. Ansari and D. LeGall, "Advanced television coding using exact reconstruction filter banks," in J. Woods (ed.), *Subband Image Coding*. Boston: Kluwer Academic Pub., 1991.

2. M. Vetterli and C. Herley, "Wavelets and filter banks: Theory and design," *IEEE Trans. on Acoustics, Speech, and Signal Proc.*, 1992.

3. M. Antonini, M. Barlaud, P. Mathieu, and I. Daubechies, "Image coding using wavelet transform," *IEEE Trans. on Image Proc.*, vol. 1, 1992, pp. 205–220.

4. A. Cohen, I. Daubechies, and J. C. Feauveau, "Biorthogonal bases of compactly supported wavelets," *Comm. Pure Appl. Math.*, vol. 45, 1992, pp. 485–560.

5. J. H. McClellan, "The design of two-dimensional digital filters by transformations," in *Proc. 7th Annual Princeton Conference on Information Sciences and Systems*, 1973, pp. 247–251.

6. D. E. Dudgeon and R. M. Mersereau, *Multidimensional Digital Signal Processing*, Englewood Cliffs, NJ: Prentice-Hall, 1983.

7. F. A. M. L. Bruekers and A. W. M. van den Enden, "New networks for perfect inversion and perfect reconstruction," *IEEE J. Sel. Areas in Commun.*, vol. 10, 1992, pp. 130–137.

8. T. G. Marshall, Jr., "Multiresolution transform matrices," *Proc. Asilomar Conf. on Signals, Syst., and Computers*, Pacific Grove, CA, 1991.

9. T. G. Marshall, Jr., "Matrix descriptions and predictive realizations for wavelets and other multiresolution operators," *Proc. Conf. on Info. Sciences and Sytems*, Princeton, NJ, 1992.

10. T. G. Marshall, Jr., "Predictive and ladder realizations of subband coders," *Proc. IEEE Workshop on Visual Signal Processing and Communications*, Raleigh, NC, 1992.

11. A. A. C. Kalker and I. A. Shah, "Ladder structures for multidimensional linear phase perfect reconstruction filter banks and wavelets," *Proc. SPIE Conf. Visual Commun. and Image Proc.*, vol. 1818, 1992, pp. 12–20.

12. T. G. Marshall, Jr., "A fast wavelet transform based upon the Euclidean algorithm," *Proc. Conf. Information Sciences and Systems*, Johns Hopkins, Baltimore, MD, 1993.

13. T. G. Marshall, Jr., "U-L block-triangular matrix and ladder realizations of subband coders," *Proc. IEEE Int. Conf. on Acoust., Speech, and Sig. Proc.*, University of Minnesota, Minneapolis, 1993, pp. III-177–III-180.

14. R. E. Van Dyck and T. G. Marshall, Jr., "Ladder realizations of fast subband/VQ coders with diamond support for color images," *Proc. IEEE Int. Symp. on Circuits and Systems*, Chicago, IL, 1993.

15. R. E. Van Dyck, N. Moayeri, T. G. Marshall, E. Simotas, and M. Chin, "Fast subband coding with ladder structures," *Proc. SPIE Conf. on Applications of Digital Image Processing*, San Diego, CA, 1993.

16. L. Tolhuizen, H. Hollman, and T. A. C. M. Kalker, "On the realizability of biorthogonal, m-dimensional two-band filter banks," *IEEE Trans. on Signal Proc.*, vol. 43, 1995, pp. 640–648.

17. A. Suslin, "On the structure of the special linear group over polynomial rings," *Math. USSR Izvestija*, vol. 11, 1977, pp. 221–238.

18. V. Belevitch, *Classical Network Theory*, San Francisco: Holden-Day, 1968.

19. P. P. Vaidyanathan, T. Q. Nguyen, Z. Doganata, and T. Saramaki, "Improved technique for design of perfect reconstruction FIR QMF banks with lossless polyphase matrices," *IEEE Trans. Acoust., Speech, Signal Processing*, vol. ASSP-37, 1989, pp. 1042–1056.

20. M. Vetterli and D. LeGall, "Perfect reconstruction FIR Filter banks: Some properties and factorizations," *IEEE Trans. Acoust., Speech, Signal Processing*, vol. ASSP-37, 1989, pp. 1057–1071.

21. T. W. Cairns, "On the fast fourier transform of finite abelian groups," *IEEE Trans. on Computers*, 1971, pp. 569–571.

22. T. G. Marshall, Jr., "The polyphase transform and its applications to block-processing and filter-bank structures," *Proc. IEEE Int. Symp. on Circuits and Systems*, Philadelphia, PA, 1987.

23. P. J. Burt and E. A. Adelson, "The Laplacian pyramid as a compact image code," *IEEE Trans. Commun.*, vol. 31, 1983, pp. 532–540.

24. J. Lim, *Two-Dimensional Signal and Image Processing*, Englewood Cliffs: Prentice Hall, 1990.

25. O. Herrmann, "On the approximation problem in nonrecursive digital filter design," *IEEE Trans. on Circuit Theory*, vol. CT-18, 1971, pp. 411–413; reprinted in *Digital Signal Processing*, L. R. Rabiner and C. M. Rader (eds.), New York: IEEE Press, 1972.

26. R. Ansari, C. Guillemot, and J. F. Kaiser, "Wavelet construction using Lagrange halfband filters," *IEEE Trans. Circuits Syst.*, vol. CAS-38, 1991, pp. 1116–1118.

27. M. G. Bellanger, *Digital Processing of Signals*, New York: Wiley-Interscience, 1984.

28. P. J. Davis, *Circulant Matrices*, New York: Wiley-Interscience, 1979.

29. C. W. Curtis and I. Reiner, *Representation Theory of Finite Groups and Associative Algebras*, New York: Wiley.

30. C. G. Cullen, *Matrices and Linear Transformations*, 2nd Ed., New York: Dover, 1990.

31. R. E. Blahut, *Fast Algorithms for Digital Signal Processing*, Reading MA: Addison-Wesley, 1984.

32. L. M. G. M. Tolhuizen, I. A. Shah, and A. A. C. M. Kalker, "On constructing regular filter banks from domain bounded polynomials," *IEEE Trans. Signal Proc.*, vol. 42, 1994.

33. E. Viscito and J. P. Allebach, "The analysis and design of multidimensional FIR perfect reconstruction filter banks for arbitrary sampling lattices," *IEEE Trans. Circuits Syst.*, vol. CAS-38, 1991, pp. 29–41.

34. D. B. H. Tay and N. G. Kingsbury, "Structures for multidimensional FIR perfect reconstruction filter banks," *Proc. IEEE Conf. on Acoust., Speech, and Sig. Proc.*, 1992, pp. IV-641–IV-644.

35. A. K. Jain, *Fundamentals of Digital Image Processing*, Englewood Cliffs: Prentice-Hall, 1989.

36. I. Daubechies, "Orthonormal bases of compactly supported wavelets II. Variations on a theme," *SIAM J. Math. Analysis*, vol. 24, 1993, pp. 499–519.

37. R. E. Van Dyck, T. G. Marshall, Jr., M. Chin, and N. Moayeri, "Wavelet video coding with ladder structures and entropy-constrained quantization," *IEEE Trans. on Circuits and Sys. for Video Tech.*, to appear.

Multidimensional Systems and Signal Processing, 8, 89–110 (1997)

On Translation Invariant Subspaces and Critically Sampled Wavelet Transforms *

STEVEN A. BENNO
Lucent Technologies, Inc., Room 1A-215, 67 Whippany Ave., Whippany, NJ 07981-0903

JOSÉ M. F. MOURA
Electrical and Computer Engineering Department, Carnegie Mellon University, Pittsburgh, PA 15213

Editor: Sankar Basu

Abstract. The discrete wavelet transform (DWT) is attractive for many reasons. Its sparse sampling grid eliminates redundancy and is very efficient. Its localized basis functions are well suited for processing non–stationary signals such as transients. On the other hand, its lack of translation invariance is a major pitfall for applications such as radar and sonar, particularly in a multipath environment where numerous signal components arrive with arbitrary delays.

The paper proposes the use of robust representations as a solution to the translation invariance problem. We measure robustness in terms of a mean square error for which we derive an expression that describes this translation error in the Zak domain. We develop an iterative algorithm in the Zak domain for designing increasingly robust representations. The result is an approach for generating multiresolution subspaces that retain most of their coefficient energy as the input signal is shifted. A typical robust subspace retains 98% of its energy, a significant improvement over more traditional wavelet representations.

Keywords: robust representations, shiftability, wavelet transform, translation invariance, Zak transform, multipath detection

1. Introduction

In many signal processing applications, we have the need to process signals that are arbitrarily shifted. For example, in radar and sonar applications, time delays and Doppler shifts occur due to a target's range and velocity, respectively. In communications, jitter creates unknown time shifts in pulse amplitude modulation schemes. The problem of arbitrary time delays is further complicated in a multipath environment. In this case, the transmitted signal travels to the receiver via different routes through the channel. In seismic signal processing, the multiple paths are caused by the transmitted signal reflecting off different stratified layers. In sonar, multipath signals result from surface and bottom reflections as well as refractive phenomena due to the nonhomogenous ocean environment caused by variations in water temperature and salinity. We discuss briefly two applications where our work is useful.

Detection in Multipath: Detecting multipath signals by a generalized maximum likelihood test (GLRT) receiver is difficult. To appreciate the complexity of the GLRT receiver, consider the following detection problem. Assume that the sent signal $g(t)$ is known and the multipath signal is produced from a linear combination of delayed replicas,

---

*   This work was partially supported by DARPA through ONR grant N00014-91-J183.

$$g_M(t) = \sum_{k=1}^{K} \alpha_k g(t - \tau_k). \tag{1}$$

The two hypotheses for the received signal $r(t)$ are

$$
\begin{aligned}
H_1 : \quad & r(t) = g_M(t) + w(t) \\
H_0 : \quad & r(t) = w(t)
\end{aligned}
$$

where, for simplicity, we assume the corrupting noise is white, Gaussian noise. The number of replicas K, the attenuation coefficients, $\{\alpha_k\}$, and the delays $\{\tau_k\}$ are all unknown. For a known signal $g(t)$, detecting the presence of $g_M(t)$ with a GLRT receiver requires calculating the maximum likelihood (ML) estimate, $g_M^\star(t)$, of $g_M(t)$ by finding those parameters, $\{K^\star, \{\alpha_k^\star\}, \{\tau_k^\star\}\}$, that minimize the distance,

$$g_M^\star(t) = \operatorname{argmin} \left\| r(t) - \sum_{k=1}^{K} \alpha_k g(t - \tau_k) \right\| \tag{2}$$

which is a prohibitively difficult multidimensional optimization problem.

In [6], we explore an alternative, simpler receiver design which is based on the geometric interpretation of the matched filter as a projection onto linear subspaces. Our approach is to approximate the optimal projection by designing subspaces whose projections are easy to compute and remain close in some sense to the true signal subspace we wish to approximate.

The present paper is concerned with the question of designing these simpler subspaces and is related to the issue of shiftability. A signal g is *shiftable* [16] if and only if

$$\forall \tau \in \mathbf{R}: \quad g(t - \tau) = \sum_{k} a_k g(t - k). \tag{3}$$

The set of functions $\{g(t - k)\}$ generates a *shiftable representation*. By using a shiftable representation, the multipath signal can be expressed as

$$g_M(t) = \sum_{k} b_k g(t - k)$$

and the ML estimate problem becomes one of finding the optimal coefficients $\{b_k^\star\}$ that minimize the distance

$$g_M^\star(t) = \operatorname*{argmin}_{b_k} \left\| r(t) - \sum_{k} b_k g(t - k) \right\|$$

which is easier than the optimization problem in (2) because $\{b_k\}$ are found by taking inner products of the received signal $r(t)$ with a dual $\tilde{g}(t)$ of the original signal $g(t)$. The solution to the original multipath detection problem is now in terms of projections onto linear subspaces spanned by the integer shifts of $g(t)$.

In general, however, (3) will not be satisfied, i.e., a function will not be shiftable. For example, it is not possible for a function to be shiftable and to have compact support [15].

We relax the hard constraint of shiftability and consider instead the design of *robust* signals for which, in (3), we have approximate equality. In other words, in this paper, we design signals for which the mean square error in representing their arbitrary shifts by its integer translates,

$$\mathcal{E}(\tau) = \|g(t-\tau) - \sum_k a_k(\tau)g(t-k)\|^2 \tag{4}$$

is acceptably small. This translation error is a measure of the representation's *robustness* to continuous translations.

For representations generated by $\{g(t-k)\}_{k\in \mathbb{Z}}$ and its biorthogonal dual $\{\tilde{g}(t-k)\}_{k\in \mathbb{Z}}$ we derive an appropriate expression for the translation error relating $\mathcal{E}(\tau)$ to the Zak transform of $|G(f)|^2$, the energy spectral density (ESD) of $g(t)$. Based on the insight provided by this expression, we propose an algorithm for reshaping the ESD of $g(t)$ in the Zak domain in order to construct a new function $g_1(t)$ such that $\mathcal{E}_{g_1}(\tau) \leq \mathcal{E}_g(\tau)$ pointwise. We demonstrate that the reshaping algorithm may be iterated to provide a sequence of functions whose corresponding translation errors decrease monotonically with each iteration, i.e., $\mathcal{E}_{g_{n+1}}(\tau) \leq \mathcal{E}_{g_n}(\tau)$.

The use of the mean square error (4) (MSE) as a design criteria for robust representations is justified for detection applications because the detection statistic is based on the energy captured by the representation. A small MSE in (4) implies that the representation will contain most of the signal's energy, regardless of the shift. As a result, the detector's performance is robust to translations. Robust representations provide a fixed subspace whose orthogonal projection is simple to compute. They can be used for detecting an arbitrary number of randomly shifted replicas of a signal.

DWT Interpolation: A second problem of interest to which our work applies is the interpolation of the discrete wavelet transform (DWT). In practice, the solution of most signal processing problems are implemented on computers which require sampling the input data. For applications that must contend with continuous delays of a signal, such as the radar and sonar problems discussed above, no discrete representation can be translation invariant to continuous shifts. In the case of the DWT, this critically sampled representation lacks translation invariance even for discrete shifts on the sampling grid. As the input is delayed in time, coefficient energy distributes itself across the time–scale plane even though the spectral content of the signal does not change. On the other hand, the DWT has many desirable features. Its sparse sampling grid avoids redundancy in the representation and the DWT is very efficient to compute. Its subband decomposition allows for a coarse to fine multiresolution analysis (MRA) and its localized basis functions are well suited for processing non–stationary signals such as transients. Robust representations provide a solution to the translation invariance problem while maintaining its other attractive attributes.

Several approaches have been developed to address the DWT's lack of translation invariance. These approaches typically revert to oversampled representations as a final solution [11] or as an intermediate step toward finding an optimal, critically sampled representation [13, 4]. Others have proposed searching through a library of shifted basis functions [7, 14]. All of these approaches forfeit the computational efficiency of the critically sampled grid and typically address only discrete shifts of an input sequence, not continuous shifts. Furthermore, some of these approaches cannot handle scenarios like the multipath case, where there are components within the input signal that are translated relative to each other. For example, the method proposed in [12] introduces a cost function μ for linear transforms $\mathcal{T}$ that is invariant to translations, i.e.,

$$\forall s \in \mathbb{Z}, \quad \mu(\mathcal{T}(\{x_k\})) = \mu(\mathcal{T}(\{x_{k-s}\})).$$

For the discrete wavelet transform, this μ–invariance approach is achieved by shifting the input signal $\{x_k\}$ to "undo", or compensate for the shift s. Their approach breaks down, however, when the signal contains multiple components that are shifted independently of each other,

$$\forall r, s \in \mathbb{Z}, \quad \mu(\mathcal{T}(\{x_k + y_k\})) \neq \mu(\mathcal{T}(\{x_{k-s} + y_{k-r}\})).$$

The problem is that their solution for μ–invariance is not *linear*. Another important distinction is that the methods proposed in [14, 12] use the lack of translation invariance as a degree of freedom for finding the optimal shift of the input signal. for a fixed wavelet basis. The distribution of the coefficients is varied by shifting the input signal in order to find that shift which leads to an optimal distribution with respect to a suitable cost function. In short, they seek the best shift for a given representation. Conversely, we seek the best representation for shifts. We view the lack of shift invariance as a nuisance, not an opportunity. Our approach is to control the redistribution of the coefficients by designing scaling functions and wavelets, independently of the input data, so that the *subspaces* they generate are translation invariant, i.e., the coefficient energy within each subband remains unchanged as the input is shifted. This neutralizes the ill effects of translation variance for detection applications because the test statistic depends on the signal's *energy* captured by the representation. If the signal of interest has its energy concentrated in one subband, a robust representation will guarantee that the decomposition of shifted versions of that signal will reside in the same subband as the unshifted version.

A robust DWT is important if one is to interpret the DWT as a time–frequency representation. Since the spectrum of the shifted signal is the same as the unshifted version, each subband should preserve its energy. This is important in detection applications for localizing the signal of interest in the time–scale plane despite the lack of translation invariance. Other applications, however, do not require this physical interpretation for the DWT. In [8], for example, entropy–based cost functions are used to design representations that have as much energy contained in as few coefficients as possible. This is useful for data compression because the coefficients with negligible energy can be zeroed. The more coefficients with negligible energy, the better the compression. In this application, there is no regard for the location of the coefficients in the time–scale plane because the decomposition is not used to analyze or interpret the input signal. This demonstrates that an optimal representation for one application may be totally inappropriate for another.

We now provide a brief summary of the paper. In Section 2, we formally present the problem of robust representations and we derive a general expression for $\mathcal{E}(\tau)$ that is derived without any assumptions on the form of the mother function $g(t)$ or its dual $\tilde{g}(t)$. We then impose an assumption on the form of the dual function so that $\tilde{g}$ is biorthogonal to g in order to simplify the general expression for the translation error. We then propose an algorithm for constructing robust functions based on this simplified expression by reshaping its energy spectral density in the Zak domain. This algorithm may be iterated as necessary to produce increasingly robust representations. In Section 5, we use the reshaping algorithm to generate robust scaling functions. As a consequence, the robust wavelet representation of a signal preserves the energy content at each scale regardless of translation.

2. Problem Statement and Definitions

As a solution to the DWT's lack of translation invariance, we propose the use of subspaces that are robust to translations of the input. In Section 5, we apply our results to the design of robust scaling functions. By doing so, we create MRAs whose subband energy is invariant to translations. Before considering the design of a robust MRA, we consider the simpler problem of designing robust representations based on time shifts only. Specifically, we are interested in designing representation subspaces $\mathcal{G} \subset L^2(\mathbb{R})$,

$$\mathcal{G} = \text{span}\{g(t - kT)\}_{k \in \mathbb{Z}}$$

such that for $h \in \mathcal{G}$, the representation captures most of the energy of the arbitrarily shifted signal $h(t - \tau)$, $\tau \in \mathbb{R}$. Without loss of generality, we assume $g(t)$ is properly scaled so that we can take $T = 1$. Furthermore, we assume g satisfies the frame condition

$$0 < A \le \sum_k |G(f + k)|^2 \le B < \infty \tag{5}$$

where A and B are the frame bounds and $G(f)$ is the Fourier transform of $g(t)$. For $h \in \mathcal{G}$,

$$\forall \tau \in \mathbb{R} \quad h(t - \tau) = \sum_k b_k g(t - \tau - k),$$

hence, the ability of the subspace $\mathcal{G}$ to include $h(t - \tau)$ depends on the ability of $\{g(t-k)\}_{k \in \mathbb{Z}}$ to represent the arbitrary shifts of the mother function $g(t - \tau)$.

Under appropriate bandlimited conditions [16, 5], it is possible for

$$\forall \tau \in \mathbb{R} \quad g(t - \tau) = \sum_k a_k(\tau) g(t - k). \tag{6}$$

The mother function g and its corresponding representation subspace $\mathcal{G}$ are said to be *shiftable* [16].

In general, however, a function is not shiftable. Other desirable properties, such as rate of decay or compact support, can preclude a function from satisfying (6) with equality. As an alternative to the hard constraint of shiftability, we introduce the *translation error*,

$$\mathcal{E}(\tau) = \|g(t - \tau) - \sum_k a_k g(t - k)\|^2, \tag{7}$$

as a measure of a signal's *robustness* to translations. In this paper, we consider the design of a mother function $g \in L^2(\mathbb{R})$ such that it generates a representation whose corresponding translation error is below an acceptable threshold, i.e., for $\epsilon > 0$,

$$\mathcal{E}(\tau) < \epsilon.$$

We define such representations as ϵ–*robust* . We then apply this notion to the design of ϵ–robust MRAs.

3.   Zak Transform and Translation Error

The main result of this section is to derive an alternate form for the translation error. This new expression relates $\mathcal{E}(\tau)$ to the energy spectral density (ESD) of the mother function via the Zak transform. We first present the definition and relevant properties of the Zak transform. For a more in–depth tutorial, see [10, 1].

3.1.   Zak Transform

The Zak transform of a function $g(x)$ is defined by

$$(Zg)(x,y) = \sum_{k} g(x+k)e^{-j2\pi ky}. \tag{8}$$

In words, for a given value of x, the Zak transform is found by taking the DFT of the sequence of integer spaced samples $\{g(x+k)\}_{k\in\mathbb{Z}}$. In this sense, the Zak transform is a mapping of a 1–dimensional function onto the 2–dimensional time–frequency plane. A similar object is also referred to as the Weil–Brezin mapping [3]. Its inverse is

$$g(x) = \int_{0}^{1} (Zg)(x,y)\, dy.$$

It will be convenient to use the following notation:

Translation: $T_a(x) = g(x-a)$.

Modulation: $(E_b g)(x) = g(x)e^{-j2\pi bx}$.

Convolution: $(g * h)(x) = \int_{-\infty}^{\infty} g(y)h(x-y)\, dy$.

The Zak transform has the following properties which can be verified by substitution into (8).

Time Shifts: $(ZT_a g)(x,y) = (Zg)(x-a,y)$
The Zak transform is *quasi–periodic* along the x–axis,

$$(ZT_n g)(x,y) = e^{-j2\pi ny}(Zg)(x,y).$$

Modulation: $(ZE_b g)(x,y) = e^{-j2\pi bx}(Zg)(x,y+b)$
The Zak transform is *periodic* along the y–axis,

$$(ZE_m g)(x,y) = (Zg)(x,y).$$

Quasi–Periodicity: Because of the Zak transform's periodicity along the y–axis and quasi–periodicity along the x–axis, the Zak transform is said to be quasi–periodic. As a result, the Zak transform is completely specified by the $[0, 1) \times [0, 1)$ unit square and the mapping $Z : L^2(\mathbb{R}) \mapsto L^2([0, 1) \times [0, 1))$ is unitary.

Convolution: The Zak transform of the convolution of two functions is the convolution of their Zak transforms along the x–axis,

$$
\begin{aligned}
Z(g * h)(x, y) &= \int_0^1 (Zg)(\eta, y)(Zh)(x - \eta, y)\, d\eta \\
&= ((Zg) \underset{x}{*} (Zh))(x, y).
\end{aligned}
$$

In the special case where $h(x)$ is an impulse train, $h(x) = \sum_k a_k \delta(x - k)$,

$$
Z(g * h)(x, y) = \left[\sum_k a_k e^{-j2\pi ky} \right] (Zg)(x, y).
$$

Multiplication: The Zak transform of the product of two functions is the convolution of their Zak transforms along the y–axis,

$$
\begin{aligned}
Z(g \cdot h)(x, y) &= \int_0^1 (Zg)(x, \eta)(Zh)(x, y - \eta)\, d\eta \\
&= ((Zg) \underset{y}{*} (Zh))(x, y).
\end{aligned}
$$

In the special case where h is periodic, i.e., $h(x) = \sum_k c_k e^{-j2\pi kx}$,

$$
Z(g \cdot h)(x, y) = h(x)(Zg)(x, y).
$$

Fourier Transform Pairs: The Zak transforms of a function g and its Fourier transform G are related by

$$
(Zg)(x, y) = e^{j2\pi xy}(ZG)(y, -x).
$$

3.2. *Unconstrained Translation Error*

We now proceed to express $\mathcal{E}(\tau)$ in a form that we can use for the design of robust representations. We derive a general expression which makes no assumptions on the mother function except that it satisfies the frame condition (5) and that its dual exists. In the following subsection, we assume a form for the dual which simplifies the expression for $\mathcal{E}(\tau)$ which we use as the basis for our algorithm.

We rewrite the translation error given by (7) in the Fourier domain

$$\mathcal{E}(\tau) = \left\| G(f)e^{-j2\pi f\tau} - \left(\sum_k a_k e^{-j2\pi fk} \right) G(f) \right\|^2. \tag{9}$$

If the integer translates $\{g(t-k)\}_{k\in\mathbb{Z}}$ are not an orthonormal basis of $\mathcal{G}$, it is well known that the coefficients $\{a_k\}$ of the representation are computed by inner products with the integer translates of a dual function $\tilde{g}$ of g, i.e.,

$$
\begin{aligned}
a_k &= \int_{-\infty}^{\infty} g(t-\tau)\overline{\tilde{g}(t-k)}\, dt \\
&= \int_{-\infty}^{\infty} G(\omega)\overline{\tilde{G}(\omega)}e^{-j2\pi\omega(\tau-k)}\, d\omega.
\end{aligned}
$$

It follows that

$$\sum_k a_k e^{-j2\pi fk} = \int_{-\infty}^{\infty} G(\omega)\overline{\tilde{G}(\omega)}e^{-j2\pi\omega\tau}\left(\sum_k e^{-j2\pi(f-\omega)k} \right) d\omega \tag{10}$$

$$= \int_{-\infty}^{\infty} G(\omega)\overline{\tilde{G}(\omega)}e^{-j2\pi\omega\tau}\left(\sum_k \delta(f-\omega+k) \right) d\omega \tag{11}$$

$$= e^{-j2\pi f\tau}\sum_k G(f+k)\overline{\tilde{G}(f+k)}e^{-j2\pi\tau k}. \tag{12}$$

In (12), the sum on the right hand side is identified as $Z_{G\overline{\tilde{G}}}(f,\tau)$, the Zak transform of the product $G(f)\overline{\tilde{G}(f)}$. Substituting (8) and (12) into (9),

$$\mathcal{E}(\tau) = \int_{-\infty}^{\infty} |G(f) - Z_{G\overline{\tilde{G}}}(f,\tau)G(f)|^2\, df \tag{13}$$

$$= \|g\|^2 + \sum_k \int_0^1 |Z_{G\overline{\tilde{G}}}(f+k,\tau)|^2|G(f+k)|^2\, df$$

$$-2\mathrm{Re}\left[\sum_k \int_0^1 Z_{G\overline{\tilde{G}}}(f+k,\tau)|G(f+k)|^2 \right] df, \tag{14}$$

where $\mathrm{Re}[H(x)] = \frac{1}{2}(H(x) + \overline{H(x)})$ is the operator which extracts the real component of a complex function. Using the quasi–periodicity property of the Zak transform,

$$\mathcal{E}(\tau) = \|g\|^2 + \int_0^1 |Z_{G\overline{\tilde{G}}}(f,\tau)|^2 \sum_k |G(f+k)|^2\, df$$

$$-2\int_0^1 Z_{G\overline{\tilde{G}}}(f,\tau)\sum_k |G(f+k)|^2 e^{j2\pi\tau k} \tag{15}$$

$$= \|g\|^2 + \int_0^1 |Z_{G\tilde{G}}(f,\tau)|^2 Z_{|G|^2}(f,0)\,df - 2\int_0^1 \overline{Z_{|G|^2}(f,\tau)}Z_{G\tilde{G}}(f,\tau)\,df \tag{16}$$

The real operator was dropped because it can be shown that its argument is real. Equation (16) expresses, via the Zak transform, the translation error in terms of the Fourier transforms of the autocorrelation function of g and of the cross–correlation of g with its dual $\tilde{g}$. In deriving (16), we made no assumptions regarding the dual except that it exists. Furthermore, we see that the Zak transform arises naturally in this error expression.

In the next subsection, we restrict the form of the dual function. This yields a new expression for $\mathcal{E}(\tau)$ which is the basis for our algorithm which reshapes a function in the Zak domain to create a new mother function that is more robust to translations. The reshaping algorithm can be iterated as necessary to obtain a function whose translation error is acceptably small. The concept of working in the Zak domain to modify signals originated in [2] in the context of multitarget radar problems.

3.3. Translation Error and Biorthogonality

Let $G(f)$ and $\tilde{G}(f)$ be the Fourier transforms of the original signal and its dual respectively. The dual is not unique. We work with the biorthogonal dual defined by its Fourier transform

$$\tilde{G}(f) = \frac{G(f)}{\sum_k |G(f+k)|^2}. \tag{17}$$

We assume that the signal $g(t)$ satisfies the frame condition (5). For this choice, $\{\tilde{g}(t-k)\}$ is *biorthogonal* to $\{g(t-k)\}$, i.e.,

$$\langle g(t-m), \tilde{g}(t-n)\rangle = \delta_{m,n}.$$

The biorthogonal dual in (17) greatly simplifies the expression for the translation error given in (16) as we show next. From the definition of the Zak transform, taking advantage of the fact that the denominator in (17) is periodic, we obtain successively for the Zak transform of the Fourier transform of the cross–correlation of g and $\tilde{g}$

$$Z_{G\tilde{G}}(f,\tau) = \sum_n G(f+n)\overline{\tilde{G}(f+n)}e^{-j2\pi n\tau} \tag{18}$$

$$= \frac{\sum_n |G(f+n)|^2 e^{-j2\pi n\tau}}{\sum_k |G(f+k)|^2} \tag{19}$$

$$= \frac{Z_{|G|^2}(f,\tau)}{Z_{|G|^2}(f,0)}. \tag{20}$$

Substituting (18) into (16) and assuming the function g to be of unit norm, $\|g\|^2 = 1$, the translation error becomes

$$\mathcal{E}(\tau) = 1 - \int_0^1 \frac{|Z_{|G|^2}(f, \tau)|^2}{Z_{|G|^2}(f, 0)}\, df. \tag{21}$$

Equation (21) shows, via the Zak transform, that the error $\mathcal{E}(\tau)$ depends only on the energy spectral density (ESD) of the signal $g(t)$. Hence, two signals with the same ESD are equally robust. Second, for a given delay τ_0, the translation error, $\mathcal{E}(\tau_0)$, depends only on the area under the Zak transform of the ESD $|Z_{|G|^2}(f, \tau_0)|^2$ (normalized by $Z_{|G|^2}(f, 0)$). In other words, the shape of each slice of the Zak of the ESD parallel to the frequency axis f does not affect $\mathcal{E}(\tau)$, only the area under it does. Equation (21) is the basis for the reshaping algorithm. Intuitively, we would like to reshape the surface $Z_{|G|^2}(f, \tau)$ so that the integrand in (21) will have more area under it for those values of τ where $\mathcal{E}(\tau)$ is small. However, we need to reshape $Z_{|G|^2}(f, \tau)$ with care in order to guarantee that the reshaped function is still the Zak of an ESD. Before we consider this algorithm, we need to translate the properties of an ESD from the Fourier domain to the Zak domain. We do this next.

3.4. Properties of ESDs in the Zak Domain

We recall that, for a frequency domain function $S_h(f)$ to be the ESD of a real signal $h \in L^2(\mathbb{R})$, it must possess the following properties:

ESD 1: Finite Energy: $\int_{-\infty}^{\infty} S_h(f)\, df < \infty.$

ESD 2: Real function: $S_h(f) = \overline{S_h(f)}.$

ESD 3: Even symmetry: $S_h(f) = S_h(-f).$

ESD 4: Non–negativity: $S_h(f) \geq 0.$

Because the Zak transform is unitary, these properties map onto properties in the Zak domain. We show that the following are their equivalent properties in the Zak domain:

ZESD 1: Finite Energy: $\int_0^1 Z_{S_h}(f, 0)\, df < \infty.$

Proof: The energy of h is given by its ESD

$$\int_{-\infty}^{\infty} S_h(f)\, df = \sum_k \int_0^1 S_h(f + k)\, df.$$

Interchanging the order of the summation and integral, it follows from the definition of the Zak transform

$$\|h\|^2 = \int_0^1 Z_{S_h}(f, 0)\, df.$$

ZESD 2: Real function: $Z_{S_h}(f, \frac{1}{2} + \tau) = \overline{Z_{S_h}(f, \frac{1}{2} - \tau)}$.

Proof: Because the Zak transform is periodic in τ and $S_h = \overline{S_h}$,

$$
\begin{aligned}
Z_{S_h}(f, \tau + \frac{1}{2}) &= Z_{S_h}(f, \tau - \frac{1}{2}) \\
&= \sum_k S_h(f + k) e^{j2\pi k(\frac{1}{2} - \tau)} \\
&= \overline{Z_{S_h}(f, \frac{1}{2} - \tau)}.
\end{aligned}
$$

ZESD 3: Real and Even symmetry: $Z_{S_h}(\frac{1}{2} + f, \tau) = e^{-j2\pi\tau} \overline{Z_{S_h}(\frac{1}{2} - f, \tau)}$.

Proof: Because of the Zak transform's quasi–periodicity in f,

$$
\begin{aligned}
Z_{S_h}(f + \frac{1}{2}, \tau) &= e^{-j2\pi f \tau} Z_{S_h}(f - \frac{1}{2}, \tau) \tag{22} \\
&= e^{-j2\pi f \tau} \sum_k S_h(f - \frac{1}{2} + k) e^{-j2\pi k\tau}. \tag{23}
\end{aligned}
$$

Let $k' = -k$. Since $S_h(f) = \overline{S_h(-f)}$,

$$
\begin{aligned}
\sum_k S_h(f - \frac{1}{2} - k) e^{j2\pi k\tau} &= \sum_k \overline{S_h(\frac{1}{2} - f + k)} e^{j2\pi k\tau} \\
&= \overline{Z_{S_h}(\frac{1}{2} - f, \tau)}.
\end{aligned}
$$

Substituting this expression into (23) yields the desired result.

ZESD 4: Real and Non–negative: $\int_0^{1/2} \text{Re}\, [Z_{S_h}(f, \tau)] \, d\tau \geq 0$.

Proof: Using the inverse Zak transform and separating the limits of integration into two regions,

$$
\begin{aligned}
S_h(f) &= \int_{-\frac{1}{2}}^0 Z_{S_h}(f, \tau) \, d\tau + \int_0^{\frac{1}{2}} Z_{S_h}(f, \tau) \, d\tau \tag{24} \\
&= \int_0^{\frac{1}{2}} [Z_{S_h}(f, -\tau) + Z_{S_h}(f, \tau)] \, d\tau. \tag{25}
\end{aligned}
$$

Because S_h is real, $Z_{S_h}(f, \tau) = \overline{Z_{S_h}(f, -\tau)}$ and the result follows.

The real and even symmetry properties *ZESD 2* and *ZESD 3* show that the Zak transform of an ESD is symmetric along the $\tau = \frac{1}{2}$ axis and the $f = \frac{1}{2}$ axis. Coupled with the fact that a signal is completely specified by its Zak transform on the unit square, we conclude that the Zak transform of an ESD need only be specified in one quadrant of the unit square.

We use these properties in the next section to develop an algorithm in the Zak domain that reshapes signals into more robust signals.

4. Reshaping Algorithm

The goal is to reshape the signal g so that the translation error of the resulting signal is decreased. We use the expression for $\mathcal{E}(\tau)$ given in (21) as the basis for this algorithm. The goal is to modify g to make $\mathcal{E}(\tau)$ as small as possible. Let

$$C(\tau) = \left[\int_0^1 \frac{|Z_{|G|^2}(f,\tau)|^2}{Z_{|G|^2}(f,0)} \, df \right]^{-1/2}. \tag{26}$$

We reshape the Zak transform $Z_{|G|^2}(f,\tau)$ as

$$Z_{\hat{G}}(f,\tau) = C(\tau) Z_{|G|^2}(f,\tau). \tag{27}$$

Since $C(\tau)$ is related to the reciprocal of the integral in (21), it is large for values of τ where the integral is small. By choosing $C(\tau)$ as in (26), the representation error is zero if $Z_{\hat{G}}(f,\tau)$ is the Zak transform of a valid ESD. Note that $C(\tau)$ is a real, even, non–negative function of τ so multiplying $Z_{|G|^2}(f,\tau)$ by $C(\tau)$ preserves the properties necessary for $Z_{\tilde{G}}$ to be real and even. There is no guarantee, however, that the non–negativity requirement, property *ZESD 4*, is satisfied.

To meet this condition, we proceed as follows. Let $\hat{G}(f)$ be the frequency domain signal defined by the Zak transform given in (27). We then define in the frequency domain a new signal by

$$G_1(f) = \sqrt{|\hat{G}(f)|} e^{j\theta_G(f)} \tag{28}$$

where $\theta_G(f)$ is the phase of the original signal $G(f)$. Note that the phase of $G_1(f)$ may be chosen arbitrarily since it does not affect the translation error. This choice of the phase makes the reshaped signal closely resemble the original signal. Finally, the reshaped signal $g_1(t)$ is obtained by taking the inverse Fourier transform of $G_1(f)$. The process of reshaping in the Zak domain, (27), and square rooting in the frequency domain, (28), may be repeated as necessary to achieve a satisfactory translation error. Next, we provide experimental results demonstrating the improved robustness gained by the reshaping algorithm is significant.

4.1. Examples

In this example, we demonstrate our results by using the Daubechies D8 scaling function as the initial function to the reshaping algorithm. The samples of the D8 scaling function are found according to [17] at a suitably high sampling rate. The discrete Zak transform [1] and numerical integration are used to approximate the expression for $\mathcal{E}(\tau)$ in (21). Figures 1 and 2 show the Zak transforms of the ESD for the original and twice iterated reshaped functions, respectively. In figure 2, notice the result of the reshaping algorithm is to make $Z_{|G_2|^2}(f,\tau)$ more independent on τ so that its area, when normalized by $Z_{|G_2|^2}(f,0)$, yields a smaller error in (21). Figure 3 is a plot of the two functions on the left and and a comparison of their corresponding translation errors on the right for each iteration. The first iteration provides a 5.0 dB improvement over the original function and an additional 5.1 dB improvement for the second iteration at $\tau = \frac{1}{2}$ for a total of 10.1 dB reduction over the original scaling function. The new representation looses only 2.5% of its energy whereas the original looses 25.5%. From the left plot of figure 3, it is particularly striking how closely the reshaped signal resembles the original function even after two iterations of the algorithm, and yet, the translation error has been reduced by an order of magnitude.

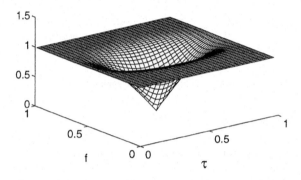

Figure 1. Zak transform of ESD of Daubechies D8 scaling function

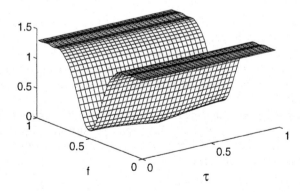

Figure 2. Zak transform of ESD of twice reshaped Daubechies D8 scaling function

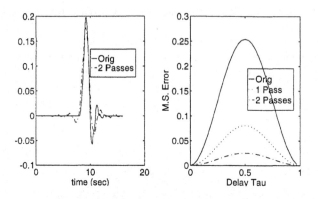

Figure 3. Daubechies D8 scaling function passed through the algorithm twice.

4.2. *Alternate Reshaping Algorithm*

While the signal resulting from the reshaping algorithm presented above is more robust than the original signal, it contains artifacts which may be improved upon. For example, if the initial $g(t)$ is of compact support, there is no guarantee that the resulting $g(t)$ will. Furthermore, square rooting of the magnitude of $\hat{G}(f)$ introduces undesirable nonsmoothness effects on zero crossings of the function. To remedy these shortcomings, the following variation is proposed as an alternative to the aforementioned algorithm.

Let $C(\tau)$ and $\hat{G}(f)$ be given as before by equations (26) and (27), respectively, but, instead of using (28), define $G_1(f)$ as

$$G_1(f) = |\hat{G}(f)|^2 e^{j\theta_G(f)} \tag{29}$$

where $\theta_G(f)$ is, as before, the phase of the original signal $G(f)$.

There are several advantages to the modified algorithm. First, by squaring the spectrum instead of taking its absolute value to enforce the non–negativity constraint of ESDs, the spectrum of G_1 is smoother, resulting in a $g_1(t)$ with side lobes with faster decay, as illustrated in figure 4.

The second advantage to the modified algorithm is that if the original $g(t)$ has compact support, then the resulting $g_1(t)$ will also have compact support. This is not guaranteed by the first reshaping algorithm. We show this by first realizing that $C(\tau)$ is a periodic function in τ. Let its Fourier series be

$$C(\tau) = \sum_k c_k e^{-j2\pi k\tau}. \tag{30}$$

By taking the inverse Zak transform, we can show that $\hat{G}(f)$ is related to the original function by

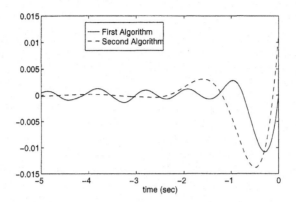

Figure 4. Comparison of sidelobes for the two reshaping algorithms.

$$\hat{G}(f) = \sum_k c_k |G(f + k)|^2$$

which is the result of convolving $|G(f)|^2$ with the impulse train $\sum_k c_k \delta(f + k)$. In the time domain, $\hat{g}(t)$ is related to the autocorrelation $R_g(t)$ of the original function multiplied by the periodic function in (30)

$$\hat{g}(t) = C(t) R_g(t).$$

Hence, the intermediate function $\hat{g}(t)$ has compact support if the original function does. From (29), $g_1(t)$ is related to the autocorrelation function of $\hat{g}(t)$

$$g_1(t) = \Theta_G(t) * R_{\hat{g}(t)}$$

where $\Theta_G(t)$ is the inverse Fourier transform of $e^{j\theta_G(f)}$. Since $g(t) = \Theta_G(t) * \mathcal{F}^{-1}[|G(f)|]$ has compact support, we conclude that $g_1(t)$ has compact support as well. Taking $\Theta_G(t) = \delta(t)$, it becomes clear that if the original function has compact support with duration D, then $g_1(t)$ will have duration $4D$. Furthermore, $g_1(t)$ is smoother than the original function because it is derived from the autocorrelation of the intermediate function $\hat{g}(t)$ which is itself derived from the autocorrelation of the original function.

In the sequel, any reference to the reshaping algorithm is limited to this alternate algorithm.

4.3. *Examples*

Figure 5 shows the resulting $g(t)$ after one single iteration of applying the algorithm to the Daubechies D8 scaling function normalized to unit energy. The reshaped function does not look drastically different from the original, but, it is much more robust. The worst case at

$\tau = \frac{1}{2}$ has been reduced by a factor of 25 times over the original function, a 14 dB reduction in translation error.

Figures 6 and 7 show the results using the D16 and the D32 scaling functions as the initial function to the reshaping algorithm, respectively. The results are very similar, with 13.4 dB and 13.5 dB improvements. As before, the reshaping algorithm can be iterated further for greater loss reduction.

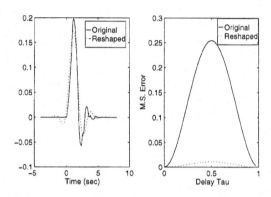

Figure 5. Original D8 and reshaped scaling functions.

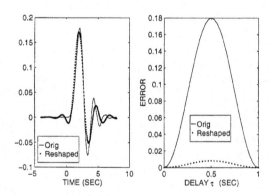

Figure 6. Original D16 and reshaped scaling functions.

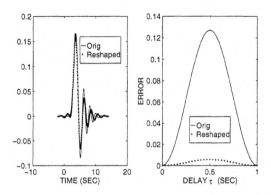

Figure 7. Original D32 and reshaped scaling functions.

4.4. ϵ–Robust Orthogonal Representations

The reshaping algorithms do not preserve orthogonality even when the original function is orthogonal. By using the orthogonalizing trick in [9], we can generate ϵ–robust orthonormal functions as follows. A non–orthogonal representation $\{g(t-k)\}_{k\in\mathbb{Z}}$ can be orthogonalized via its Fourier transform,

$$G'(f) = \frac{G(f)}{\sqrt{\sum_k |G(f+k)|^2}}. \tag{31}$$

Equation (31) is well behaved since we have assumed g satisfies the frame condition (5). Equation (31) may be applied after the reshaping algorithm has been iterated sufficiently to ensure that the resulting orthonormal representation is ϵ–robust .

Figure 8 is the result of orthogonalizing the reshaped D8 scaling function after one iteration of the algorithm. The plot on the right shows the translation error for the original D8, for the orthogonalized reshaped function, and for the 1–pass non–orthogonalized version from the previous example for comparison. The orthogonalized reshaped function is 2.9 dB down from the original D8 at $\tau = \frac{1}{2}$, but is 2.1 dB higher than the non–orthogonalized function. This is due, in part, to the fact that the redundancy in the non–orthogonal representation helps to reduce $\mathcal{E}(\tau)$ [5]. Nonetheless, the algorithm can be iterated to construct ϵ–robust orthogonal representations.

5. ϵ–Robust Scaling Functions

Our goal in this section is to design robust MRAs. The reshaping algorithm preserves certain properties of the initial function. For example, we have shown in Section 4.2 that

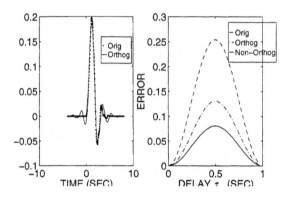

Figure 8. Original D8 and orthogonalized reshaped scaling functions.

the resulting function has compact support if the initial function has compact support. Conversely, orthogonality is not preserved by the reshaping algorithm but can be recovered via the orthogonalizing "trick" as in Section 4.4. Strictly speaking, initializing the reshaping algorithm with a scaling function does not guarantee the resulting function will satisfy the 2–scale equation. Experimental results, however, demonstrate that the reshaped scaling function is "almost" a scaling function, from which we find a valid robust scaling function.

In order to determine if a function is indeed a scaling function, we derive the Zak domain equivalent of the 2–scale equation which provides a means for identifying valid scaling functions and for obtaining their underlying 2–scale coefficients.

5.1.   Zak Domain and 2–Scale Equation

Define $\phi_j(t) = \phi(2^j t)$. If $\phi(t)$ is a valid scaling function, it must satisfy the 2–scale equation

$$\phi_{-1}(t) = \sum_k h_k \phi_0(t - k).$$

Taking the Zak transform of the 2–scale equation yields

$$Z_{\phi_{-1}}(t, f) = H(f) Z_{\phi_0}(t, f) \tag{32}$$

where $H(f) = \sum_k h_k e^{-j2\pi f k}$. This is the Zak domain equivalent of the 2–scale equation. The significance of (32) lies in the observation that Z_{ϕ_0} and $Z_{\phi_{-1}}$ are both non–trivial functions of f and t, yet, their ratio is a periodic function of f and independent of t. The periodicity of $H(f)$ comes free because of the Zak transform's properties. The ratio's independence on t, however, is a non–trivial consequence of $\phi(t)$ satisfying the 2–scale equation. We will use the ratio $\frac{Z_{\phi-1}(t,f)}{Z_{\phi 0}(t,f)}$ [1] to test if a signal g does in fact satisfy the 2–scale equation.

5.2. Reshaped Scaling Functions

The overwhelming question is whether or not the reshaping algorithm produces valid scaling functions. In fact, it can be shown that, in general, the algorithm does not, even though the initial function does satisfy the 2–scale equation. We have found experimentally, however, that the reshaped scaling function is, itself, "almost" a scaling function.

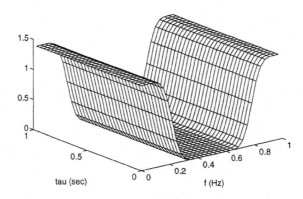

Figure 9. Ratio of Zaks of reshaped function $\frac{Z_{g-1}(t,f)}{Z_{g0}(t,f)}$.

We ran the algorithm first with $\phi(t)$ as the starting function, then with $\phi(t/2)$ to get $g(t)$ and $g(t/2)$, respectively. Next, we took the ratio of their Zak transforms to identify if $g(t)$ satisfies the 2–scale equation. We have used the D8 scaling function once again to demonstrate this procedure and have plotted the ratio of Zak transforms in figure 9. At first glance, $\frac{Z_{g-1}(t,f)}{Z_{g0}(t,f)}$ appears to be independent of t but closer inspection reveals otherwise.

Figure 10 is a closeup plot of figure 9 as a function of f. Each line corresponds to a different value of t. If the ratio were independent of t, the different lines in figure 10 would not be discernible. Nonetheless, $g(t)$ is an excellent approximation to a scaling function and the coefficients $\{h_k\}$ of the 2–scale equation can be gotten by several different ways. One way is by taking $H(f)$ to be any slice of the ratio in figure 9 parallel to the f axis. Alternately, $H(f)$ can be taken as an average of the different slices,

$$H(f) = \int_0^1 \frac{Z_{g-1}(t, f)}{Z_{g0}(t, f)} \, dt.$$

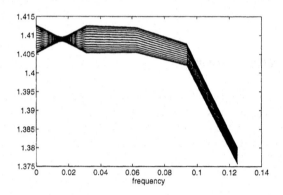

Figure 10. Closeup of ratio of Zak transforms.

The coefficients for this example were gotten by taking the inverse DFT of a single slice of the ratio and are plotted in figure 11. As a measure of the quality of the approximation, consider the mean square error

$$\|g_{-1}(t) - \sum_k h_k g_0(t - k)\|^2 = 3.5 \times 10^{-6}. \tag{33}$$

Because the duration of the reshaped function is four times the duration of the original function, the number of coefficients also increase by a factor of four. Figure 11 is a plot of the coefficients generated by this example.

Similar results have been found for all of the Daubechies scaling functions tested. As was the case with the D8 scaling function in (33), the reshaped D16 function approximately satisfies the 2–scale equation with a mean square error of 1.3×10^{-6}.

Repeated iterations of the reshaping algorithm only improves the approximation to a scaling function. This occurs because the algorithm tends to drive $Z_{|G|^2}(f, \tau)$ independent of τ, which, from (21), is a sufficient condition for driving $\mathcal{E}(\tau)$ toward zero. As a side effect, the ratio $\frac{Z_{g_{-1}}(t,f)}{Z_{g_0}(t,f)}$ also becomes increasingly independent of t.

6. Conclusion

We have addressed the issue of subspace representations that are robust to translations of the input signal. We measured robustness by the translation error, i.e., the mean square error in representing a signal and its arbitrary delays in terms of the integer shifts of a mother function. We expressed the translation error by the Zak transform of the mother function and its dual. We simplified this expression by assuming a form for the dual which makes it biorthogonal to the mother function. Based on this simplified expression, we designed an

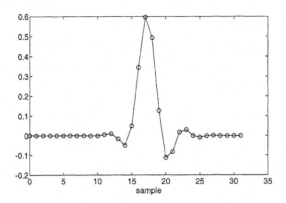

Figure 11. 2–scale coefficients of reshaped D8 signal.

algorithm for generating robust representations by reshaping the energy spectral density of the mother function in the Zak domain. This reshaping algorithm can be iterated to produce increasingly robust subspaces.

We applied the notion of robust representations to critically sampled dyadic wavelet transforms as a solution to the DWT's lack of translation invariance. Our goal is to maintain the efficiency and sparseness of the critically sampled dyadic grid but to generate a multiresolution analysis whose subband energy is mostly invariant to translations of the input signal. We have applied our results to the design of robust MRAs.

Our results demonstrate the design of scaling functions that retain a minimum of 98% of its coefficient energy as the input is translated off the grid. This represents a significant improvement, 14dB, over the traditional Daubechies D8 scaling function. Similar improvements were found for different scaling functions.

Manuscript by Benno and Moura

as per your request, here is the .bbl file. the order of the biblio entries may not be the one you had on your galley proofs.

Notes

1. $Z_{\phi_0}(t, f)$ may have zeros. However, for functions that satisfy the 2–scale equation, the ratio is bounded and well behaved since it represents the DFT of the 2–scale equation

References

1. L. Auslander, I. Gertner, and R. Tolimieri. The discrete Zak transform application to time–frequency analysis and synthesis of non–stationary signals. *IEEE Trans. Sig. Proc.*, 39(4):825–835, April 1991.

2. Louis Auslander and Frank Geshwind. Multi–target ambiguity functions. In José M. F. Moura and Isabel M. G. Lourtie, editors, *Acoustic Signal Processing for Ocean Exploration*, pages 473–489. NATO ASI, Kluwer Academic Publishers, July 1992.

3. Louis Auslander and R. Tolimieri. Radar ambiguity functions and group theory. *SIAM J. Math. Anal.*, 16(3):577–601, May 1985.

4. Feng Bao and Nurgün Erdöl. The optimal wavelet transform and translation invariance. In *ICASSP*, pages III–13–16, April 1994.

5. Steven A. Benno and José M. F. Moura. Scaling functions optimally robust to translations. submitted for publication.

6. Steven A. Benno and José M. F. Moura. Shiftable representations and multipath processing. In *Proc. IEEE-SP Int'l. Symp. on Time–Freq. and Time–Scale.Anal.*, pages 397–400, October 1994.

7. I. Cohen, S. Raz, and D. Malah. Shift invariant wavelet packet bases. In *ICASSP*, pages II–1081–1084, May 1995.

8. Ronald R. Coifman, Yves Meyer, Steven Quake, and M. Victor Wickerhauser. Signal processing and compression with wavelet packets. preprint, April 1990.

9. Ingrid Daubechies. *Ten Lectures on Wavelets*. Regional Conference Series in Applied Mathematics. Society for Industrial and Applied Mathematics, Philadelphia, Pennsylvania, 1992.

10. Augustus J. E. M. Janssen. The Zak transform: A signal transform for sampled time–continuous signals. *Philips Journal of Research*, 43(1):23–69, 1988.

11. R. Kronland-Martinet, J. Morlet, and A. Grossman. Analysis of sound patterns through wavelet transforms. *Int. J. Pattern Rec. Art. Intell*, 1(2):97–126, 1987.

12. Jie Liang and Thomas W. Parks. A two-dimensional translation invariant wavelet transform with applications. In *Proc. 1st IEEE Int'l. Conf. on Image Processing*, pages 66–70, November 1994.

13. Stephane G. Mallat. Zero–crossings of a wavelet transform. *IEEE Trans. Sig. Proc.*, 37(4):1019–1033, July 1991.

14. Stephen Del Marco, Peter Heller, and John Weiss. An M–band, 2–dimensional translation invariant wavelet transform and applications. In *ICASSP*, pages II–1077–1080, May 1995.

15. J. E. Odegard, R. A. Gopinath, and C. S. Burrus. Optimal wavelets for signal decomposition and the existence of scale limited signals. In *ICASSP*, pages IV–597–600, October 1992.

16. Eero P. Simoncelli, William T. Freeman, Edward H. Adelson, and David J. Heeger. Shiftable multiscale transforms. *IEEE Trans. Info. Theory*, 38(2):587–607, March 1992.

17. Gilbert Strang. Wavelets and dilation equations: A brief introduction. *SIAM Review*, 31(4):614–627, December 1989.

Multidimensional Systems and Signal Processing, 8, 111–128 (1997)

Low Bit-Rate Design Considerations for Wavelet-Based Image Coding

MICHAEL LIGHTSTONE
Chromatic Research Corporation, 615 Tasiman Drive, Sunnyvale, CA 94089-1707

ERIC MAJANI
Jet Propulsion Laboratory, California Institute of Technology, 4800 Oak Grove Drive, Pasadena, CA 91109

SANJIT K. MITRA
*Center for Information Processing Research, Department of Electrical and Computer Engineering,
University of California, Santa Barbara 93106*

Abstract. Biorthogonal and orthogonal filter pairs derived from the family of *binomial product filters* are considered for wavelet transform implementation with the goal of high performance lossy image compression. Using experimental rate-distortion performance as the final measure of comparison, a number of new and existing filters are presented with excellent image coding capabilities. In addition, numerous filter attributes such as orthonormality, transition band sharpness, coding gain, low-band reconstruction error, regularity, and vanishing moments are assessed to determine their importance with regards to the fidelity of the decoded images. While image data compression is specifically addressed, many of the proposed techniques are applicable to other coding applications.

Key Words: wavelets, image coding, filter design

1. Introduction

An important objective in wavelet-based data compression is the determination of successful design criteria for selecting an appropriate set of basis functions. Because of the hierarchical implementation of the discrete wavelet transform, all basis functions can be uniquely determined from a single prototype, two-band perfect reconstruction filter bank as shown in Figure 1. Consequently, the problem of selecting effective basis functions and appropriate design criteria can be reduced to a search over all orthogonal and biorthogonal filter banks that satisfy the now well-known perfect reconstruction conditions [1] given by

$$H_0(z)G_0(z) + H_1(z)G_1(z) = 2, \tag{1}$$

and

$$H_0(-z)G_0(z) - H_1(-z)G_1(z) = 0, \tag{2}$$

where $H_0(z)$, $H_1(z)$, $G_0(z)$ and $G_1(z)$ refer to the lowpass and highpass filters for the analysis and synthesis sections, respectively. If we define the *product filter* as $P(z) = H_0(z)G_0(z)$ and the analysis and synthesis highpass filters as $G_0(z) = H_1(-z)$ and $G_1(z) = H_0(-z)$, then the perfect reconstruction equations simplify to the following condition:

$$P(z) + P(-z) = 2. \tag{3}$$

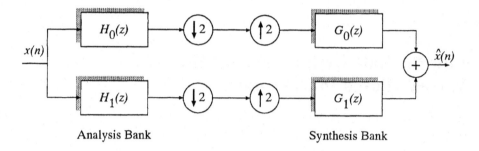

Figure 1. Analysis and synthesis sections for maximally decimated two-band filter bank.

If this constraint is satisfied, we say that the product filter is *valid* and the analysis and synthesis lowpass filters, $H_0(z)$ and $G_0(z)$, are *complementary*. Because of the nature of $P(z)$ [1], it is straightforward to show that a valid symmetric product filter $P(z)$ can be factored into either symmetric (linear phase) biorthogonal filters $H_0(z)$ and $G_0(z)$, or orthonormal filters $H_0(z)$ and $G_0(z) = H_0(z^{-1})$ [2]. Unfortunately, an exhaustive search over all valid product filters and all possible factorizations to determine the "best" subband filters is hindered by the infinite number of $P(z)$ that satisfy Eq. (3). To circumvent this difficulty, past work has focused on the design of filters that emphasize quantifiable attributes such as frequency selectivity as measured by the minimax proximity of the filter's frequency response to an ideal brick wall filter [3], [4], [5], [6] or the smoothness of the scaling and wavelet functions as measured by regularity [7]. The former quality offers promising theoretical results for subband coding if the frequency responses are ideal [8], while the latter is consistent with the smoothness typically observed in natural images and the intuitively desirable objective of retaining that smoothness after quantization. Of course, the relative merit of other filter characteristics has been debated, including orthonormality, coding gain, and vanishing moments [9], [10].

In this paper, to surmount the complexity obstacle, we limit ourselves to a family of product filters introduced by Daubechies in [11] from which a number of successful orthogonal and linear phase biorthogonal filters for image coding have already been derived. Because of this past success and other properties discussed in Section 2, we exhaustively generate a set of suitable candidate filters from all possible orthonormal and linear phase factorizations of this family up to order 26 with the goal of high performance lossy image compression. In Section 3, we define some new and existing filter bank attributes that include degree of orthonormality, transition band sharpness, coding gain, low-band reconstruction error, regularity, and number of vanishing moments. Accordingly, we assess the relative importance of each of these metrics with respect to the fidelity of the decoded image in Section 4. We stress that rather than optimizing filters for an abstract performance measure or simple quantization scheme, we focus on the search for filters and filter characteristics that maximize correlation with experimental rate-distortion performance using a state-of-the-art image codec. To our knowledge, this paper represents one of the most comprehensive experimental evaluations of its kind.

Table 1. Coefficients of $C_n(z)$ for $n = 2, 4, \ldots, 14$.

n	2	4	6	8	10	12	14
$2(n-1)$	2	6	10	14	18	22	26
c_0	1	4	38	208	5,018	32,216	430,908
$-c_{\pm 1}$		1	18	131	3,650	25,374	356,132
$c_{\pm 2}$			3	40	1,520	12,768	203,161
$-c_{\pm 3}$				5	350	4,067	79,674
$c_{\pm 4}$					35	756	20,706
$-c_{\pm 5}$						63	3,234
$c_{\pm 6}$							231

2. Binomial Product Filters

In [11], Daubechies presents a family of product filters of order $2(n-1)$ (with n even) such that

$$P_n(z) = (1 + z^{-1})^n \cdot C_n(z), \tag{4}$$

which we will refer to as the family of *binomial* product filters since they can be derived by finding the lowest order filters $C_n(z)$ which are complementary to the binomial filters $(1+z^{-1})^n$. This family of filters is important in that it has the maximum number of vanishing moments for a given product filter length. Moreover, a number of famous filter banks are derived from factorizations of this particular family of product filters. For example, the 4-tap minimum-phase Daubechies orthonormal filter is a factorization of $P_4(z)$ [11], as is the 5/3[1] biorthogonal filter proposed by Le Gall [12]. More generally, the n-tap Daubechies orthonormal filters are minimum-phase factorizations of $P_n(z)$. The coefficients of $C_n(z)$ for $n = 2, 4, \ldots, 14$, are shown in Table 1 and were computed using the formula presented in [7].

In [7], Daubechies examines some linear phase biorthogonal factorizations of the binomial product filter family, concentrating on odd-length filters with least disparate lengths, i.e. 9/7 and 11/9 filters, and on families of filters complementary to the binomial filters of orders 2, 3 and 4. Interestingly, many of these factorizations, such as the 5/3 filter ($h_0 = [-1, 2, 6, 2, -1]$ and $g_0 = [1, 2, 1]$) have been used successfully in numerous wavelet-based compression schemes. For example, the 5/3 filter has demonstrated remarkable coding performance at a very small price in complexity [13], [14], [15]. The 9/7 filter presented by Daubechies in [7] has enjoyed similar success [14], [15], [16], [10]. Image coding experiments in [10] were so successful that the FBI approved the filter pair for its fingerprint compression project [17].

The objective in this paper is to consider other possible factorizations of the binomial product filters and determine which of these also yield superior data compression performance. Our approach to the linear phase biorthogonal factorization of binomial product filters is based on the following observations and assertions:

- while the 5/3 and 9/7 filters are known examples of good filters for data compression, there is no reason to limit the search to filters of the type $2m+1/2m-1$,

- while linear phase biorthogonal filters cannot be orthonormal (except for the Haar filter), it is desirable that they be approximately orthonormal, particularly for uniform quantization,

- a sufficiently high stopband attenuation is required to minimize the amount of aliasing in the subbands, and

- filter bank coding gain provides an indication of actual performance on real image data in the fine quantization limit.

Based on these ideas, we consider *all* possible linear phase factorizations of the binomial product filters into biorthogonal filter pairs (up to order 26), and from this pool select an initial set of linear phase biorthogonal filter pairs based on a set of quantifiable filter attributes defined as stopband attenuation, degree of orthogonality, and coding gain. Filters in this pool are then evaluated by comparing their coding performance for real image data using an efficient wavelet-transform-based image coding scheme.

3. Filter Selection

While ideally we would like to compare experimental rate-distortion curves for every possible linear phase biorthogonal filter pair using real image data, the large number of possible binomial product filter factorizations makes a direct, brute-force approach very computationally expensive—especially if each filter is to code more than one image at a large number of bit rates. Instead, we limit our search to those filters that exemplify a minimal amount of stopband attenuation, orthonormality, and coding gain. While these characteristics alone may not uniquely identify the best filters, it is likely that the optimal filter exhibits some degree of these attributes. We make this assertion based on observing the nature of those filters which have provided us with the best performance in previous experiments [16]. Comparisons employing real image data are then made using these filter characteristics as indicators of potential success.

3.1. Design Criteria

We now proceed to describe the three filter characteristics (orthonormality, transition band energy, and coding gain) which are used to determine our initial pool of filters. Later we add some other metrics (low-band reconstruction error, regularity, and vanishing moments) for the purpose of comparison and evaluation. In order to obtain a measure of orthogonality, we first observe that

$$|H_0(w)|^2 + |H_1(w)|^2 = 2, \quad 0 \leq \omega \leq \pi, \tag{5}$$

for orthonormal filters—keeping in mind that $H_1(z) = G_0(-z)$. Accordingly, we define the filter $O(z) = H_0(z)H_0(z^{-1}) + H_1(z)H_1(z^{-1})$ for the more general biorthogonal case, and use its frequency response $O(w)$, to obtain a measure of the deviation from orthonormality which is given by:

$$ON = \int_0^\pi (2 - O(w))^2 \cdot dw. \tag{6}$$

As such, the measure ON corresponds to the unweighted average squared deviation from an all pass filter response (assuming it is normalized by π). It is easy to see that the measure is greater than or equal to zero for all filter pairs and is equal to zero if and only if the filters are orthonormal.

Next, in order to measure stopband attenuation, we define the filter bank transition band energy as:

$$TBE = \int_0^\pi |H_0(w)H_1(w)|^2 dw. \tag{7}$$

This function represents, in some sense, a measure of the deviation from an ideal lowpass and highpass filter pair. If the overlap between the filters is zero, which is only possible for ideal "brickwall" filters (assuming the filter bank is perfect reconstruction), then TBE is zero. Like ON, the measure is greater than or equal to zero for all filter pairs. Moreover, by employing Parseval's relation [18], both integrals in Eqs. (6) and (7) can be computed exactly using the filter coefficients of $H_0(z)$ and $G_0(z)$.

For coding gain we employ an expression derived by Katto and Yasuda in [9] that is valid for non-unitary subband transforms. Given an N band subband decomposition, define the coefficients of the k-th analysis and synthesis filter by $h_k(i), k = 1, \ldots, N$, and $g_k(j), k = 1, \ldots, N$, respectively. Assuming that we model the source using a one-dimensional Markov model with a correlation factor ρ, the resulting expression for the coding gain is given by

$$CG(\rho) = \frac{1}{\Pi_{k=1}^N (A_k B_k)^{\alpha_k}},$$

where

$$A_k = \sum_i \sum_j h_k(i) h_k(j) \rho^{|j-i|},$$

$$B_k = \sum_i g_k(i)^2,$$

and α_k is the subsampling ratio for the k-th filter (i.e. $\alpha_k = 1/N$ for a uniform subband decomposition). Note that for tree-structured filter banks, CG is a function of the filters H_0 and G_0, the first-order correlation factor ρ, and the tree-structure, itself. In addition to the work by Katto and Yasuda, other authors have independently derived an expression for the coding gain in terms of the power spectral density of an arbitrary wide-sense stationary process, and have designed filters to maximize this criterion using autoregressive speech models [19], [20].

For our case, we compute coding gain using a five level octave-band decomposition since experimentally this number of levels has provided the best performance for the largest number of images. Intuitively, this expression may appear to be the best indicator of performance with respect to image coding as it directly measures the filter bank's coding efficiency. However, it is important to note that the derivation of this expression is based on a number of assumptions in addition to that of a first-order Markov model. For example, as with most derivations of coding gain [21], the expression relies on the invocation of a high resolution asymptotic rate-distortion relationship that is not necessarily valid for low bit-rate image coding. Furthermore, Katto and Yasuda also assume the quantization errors are uncorrelated—an assumption not necessary in the derivation of coding gain for orthogonal filter banks and transforms. Finally, the measure is for a one-dimensional source, and images are obviously two-dimensional.

Another metric that we make use of to ascertain the coding efficiency of an arbitrary hierarchical filter bank is the low-band reconstruction error or $LBRE$. This attribute measures the total squared-error induced from the application of a forward and reverse N-level hierarchical transform to an ideal step-function—assuming only the lowest frequency subband is retained, i.e. all of the wavelet coefficients in the higher frequency or detail subbands are set to zero. A measure similar to this was introduced in [22] with the intent of quantifying the overshoot or ringing in the reconstructed signal rather than the total squared-error. In both cases, the intuition behind the potential benefit of this metric is rooted in the spectral nature of digital images and the way they are traditionally coded. For example, because more often than not the vast majority of energy in an image is captured in the lowest frequency subband, image coding systems typically render these coefficients with very high precision—even at low bit rates—while much of the detail information is quantized to zero. Hence, the crudeness of the $LBRE$ measure can be seen to approximate the behavior of a real image codec for an ideal step function [22].

Finally, because of their prevalence in the wavelet literature and the uncertainty as to their ultimate significance with respect to image coding [2], [10], we include the measures of regularity and vanishing moments for comparison. For our experiments REG_A, REG_S, VAN_A, VAN_S refer to the Hölder regularity of the analysis and synthesis scaling functions and the number of vanishing moments in the analysis or synthesis wavelets, respectively. Although various techniques exist for computing REG, we use the method in [23] due to the simplicity of its implementation. Note that VAN_A and VAN_S can be determined by simply computing the number of zeros occurring at $z = -1$ for $G_0(z)$ or $H_0(z)$, respectively. Finally, we denote by CF the center frequency of the analysis filter bank, i.e. the value of w at which the frequency responses of $H_0(z)$ and $H_1(z)$ are equal.

3.2. Biorthogonal Factorization

The first step in the linear phase biorthogonal factorization of the binomial product filter $P_n(z)$ is the resolving of $C_n(z)$ into elementary symmetric factors whose roots belong to one or more of the following three groups: a complex conjugate pair on the unit circle (type I), a real root and its inverse (type II), a complex conjugate pair, and their inverses (type III). Since $C_n(z)$ has no zeros on the unit circle, its roots are of the type II and III.

If $s(n)$ is the number of elementary symmetric factors of $C_n(z)$, then the total number of unique factorizations can be shown to be $(n+1) \cdot 2^{s(n)}$. For our experiments, factorizations were performed for even values of n up to 14. The first selection of filters was made by imposing an orthonormality measure ON less than 0.5, and restricting the TBE measure to be small with respect to other factorizations obtained from the same product filter. These filter pairs were then compared to those that maximized coding gain using a first-order Markov model with $\rho = 0.95$. The four best filter pairs for each n are shown in Table 2 along with their respective TBE, ON, CG, and CF. Additionally, for each n, the filter properties of Daubechies filters derived from an orthonormal factorization of the binomial product filter are provided for comparison. Note that except for $n = 10$, the coding gain achieved from the best linear phase biorthogonal factorization is superior to that of the corresponding orthonormal, Daubechies factorization. In fact, the coding gain for two of the biorthogonal filter pairs exceeds 9.92 dB (for $\rho = 0.95$) which is the theoretical maximum achievable by any orthonormal filter in a hierarchical transform [24], [25]. Interestingly, the 17/11 filter outperforms all orthogonal filters for $\rho > 0.5$ [24].

4. Coding Experiments

4.1. Algorithm Description

For experimental assessment of filter performance, a wavelet transform quantization scheme based on the embedded zero-tree method was selected for implementation [26], [27]. This approach was chosen for two important reasons. First, the method effectively exploits statistical redundancy in the wavelet transform, providing some of the best rate-distortion performance of any known image codec, and secondly, the method enables the user to precisely specify the rate at which a given image should be coded, thus, facilitating very fair comparisons between the performance of different filters. It should be noted that rates quoted in this paper correspond to real compressed bit streams using an arithmetic lossless coder.

In our first experiment, images to be coded are decomposed using a five level wavelet transform that employs circular convolution to extend the finite sequences beyond the borders. This approach is implemented so that the linear phase biorthogonal filters can be compared to the non-symmetric, orthonormal Daubechies filters. However, it should be emphasized that for both the linear phase biorthogonal and orthonormal filters alternative methods exist that still retain perfect reconstruction, including time-varying filter banks for orthonormal filters [28] and symmetric extension for the linear phase biorthogonal case [29], [30]. Both of these approaches have the potential to improve PSNR performance and reduce ringing artifacts at image boundaries. For symmetric extension, our results indicate that on average, 0.25 dB improvement in PSNR for rates between 0.1 and 1.00 bpp is possible for the linear phase biorthogonal filters with the greatest degree of improvement occurring for rates less than 0.5 bpp.

Finally, all lowpass filters are normalized so that the coefficients of their respective impulse responses sum to $\sqrt{2}$. For orthogonal filters, this restriction results in an orthonormal transformation.

Table 2. Best linear phase biorthogonal analysis filters derived from factorization of the binomial product filters using transition band energy, degree of orthonormality, and coding gain for selection. Orthonormal Daubechies filters derived from the same order product filter are also shown along with the center frequency of each filter pair. Note "Dn" corresponds to the orthonormal Daubechies filter of length n.

n	filter	CG (dB)	TBE	ON	CF	n	filter	CG (dB)	TBE	ON	CF
2	D2	8.243	0.707	0.000	0.500	2	1/3	7.910	1.225	0.385	1.000
4	5/3	9.587	0.625	0.131	0.636	10	10/10	9.659	0.477	0.242	0.602
	2/6	9.586	0.650	0.019	0.500		9/11	9.643	0.520	0.071	0.481
	6/2	7.806	0.650	0.019	0.500		17/3	9.485	0.500	0.128	0.573
	3/5	7.302	0.625	0.131	0.364		2/18	9.464	0.648	0.061	0.500
	D4	9.286	0.600	0.000	0.500		D10	9.720	0.477	0.000	0.500
6	9/3	9.616	0.543	0.121	0.602	12	10/14	9.827	0.461	0.132	0.578
	2/10	9.601	0.644	0.038	0.500		13/11	9.818	0.469	0.019	0.504
	8/4	8.381	0.627	0.547	0.694		6/18	9.781	0.507	0.110	0.579
	3/9	8.398	0.543	0.121	0.398		17/7	9.780	0.458	0.008	0.513
	D6	9.543	0.542	0.000	0.500		D12	9.749	0.456	0.000	0.500
8	6/10	9.876	0.518	0.054	0.563	14	10/18	9.923	0.454	0.046	0.547
	9/7	9.870	0.521	0.018	0.540		17/11	9.922	0.441	0.014	0.530
	13/3	9.553	0.514	0.123	0.584		11/17	9.764	0.442	0.168	0.555
	2/14	9.534	0.645	0.051	0.500		16/12	9.739	0.488	0.070	0.522
	D8	9.653	0.505	0.000	0.500		D14	9.775	0.439	0.000	0.500

4.2. Experimental Results

4.2.1. Filter Comparisons

For the purposes of comparison, the luminance components of the "lena" and "barbara" were coded at a wide range of bit rates using each filter found in Table 2 and the quantization strategy described in the previous section. The "lena" image was chosen because it represents the standard test image used in the literature. In contrast, the "barbara" image contains significantly more high frequency content due to the presence of very fine textures which make it a more challenging test image that is less easily modeled as a first-order Markov process. All of the filters were designed from orthogonal and linear phase biorthogonal factorizations of varying order, binomial product filters $P_n(z)$. In order to assess filter efficiency, the fidelity of the reconstructed image after quantization was measured using the peak signal-to-noise ratio (PSNR), defined as $PSNR = 10\log_{10}(255^2/MSE)$ where MSE denotes the mean-squared error. The outcomes of these rate-vs.-PSNR evaluations are summarized in Tables 3 and 4. Specifically, the filters with the best performance for a given $P_n(z)$ are listed in Table 3, and the filter rankings in terms of average PSNR performance for the "lena" and "barbara" images without respect to filter length are provided in Table 4. Finally, the actual rate-vs.-PSNR curves for all of the filters, though not catalogued in this paper, can be found in [15].

Table 3. Lengths of filters with best average PSNR performance for "lena" and "barbara"

n	2	4	6	8	10	12	14
filter	2/2	2/6	2/10	9/7, 6/10	9/11	13/11	17/11

Table 4. Ranking of filters with respect to average PSNR performance for "lena" and "barbara" images.

n	filter	rank			n	filter	rank		
		lena	barb	both			lena	barb	both
4	5/3	20	26	25	10	10/10	18	23	22
	2/6	13	18	17		9/11	8	7	7
	6/2	29	29	29		17/3	24	24	24
	3/5	30	30	30		2/18	14	16	15
	D4	26	25	26		D10	15	9	13
6	9/3	17	22	21	12	10/14	9	13	9
	2/10	10	15	11		13/11	3	1	2
	8/4	27	28	28		6/18	11	12	10
	3/9	28	27	27		17/7	6	5	5
	D6	25	19	19		D12	16	10	14
8	6/10	7	11	8	14	10/18	4	4	4
	9/7	2	6	6		17/11	1	2	1
	13/3	21	21	23		11/17	22	20	20
	2/14	12	14	12		16/12	5	3	3
	D8	23	17	18		D14	19	8	16

4.2.2. Filter Attributes

Using these results, we can make some important observations about the filter characteristics defined in Section 3 by generating scatter plots that compare each of these characteristics with actual coding performance. These experiments have been performed for all of the filters in Table 3 at fixed rates of 0.1, 0.25, 0.5, 0.75, and 1.0 bits per pixel (bpp). We also include the filters in Table 3 with the analysis and synthesis sections switched so that the effects of regularity and vanishing moments on the analysis and synthesis sections can be better assessed. For comparison, we use an additional measure that corresponds to the average PSNR computed over the aforementioned rates (0.1, 0.25, 0.5, 0.75, and 1.0 bpp) for both the "lena" and "barbara" images. Results for this average performance are provided in Figures 3 and 4. Note that the orthonormal filters, the odd-length biorthogonal filters, and even-length biorthogonal filters are distinguished by using o's, x's, and +'s, respectively, for labeling. These curves also include a least-squares, first-order polynomial fit to the data. To more numerically quantify the predictive capability of each filter-attribute with respect to experimental coding performance, we provide the correlation coefficients between the value of each filter-attribute and the average PSNR's of the quantized "lena" and "barbara" images in Table 6. In addition to tabulating these results for the set of all

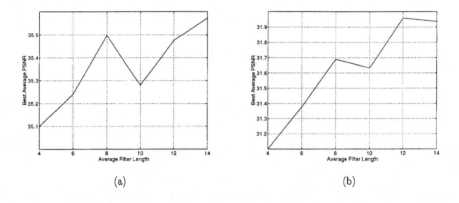

Figure 2. Best average PSNR as a function of average filter length for the (a) "lena" and (b) "barbara" images.

filters, we provide separate correlation coefficients for the odd-length biorthogonal filters, the even-length biorthogonal filters, and the orthonormal Daubechies filters.

As we analyze the data in Table 3, we first observe that for every even value of n equal to or below 14, the best filters at all rates—as measured by the PSNR of the decoded images—are linear phase biorthogonal rather than orthonormal. Secondly, among the measures of filter performance defined in Section 3, the one-dimensional coding gain expression provides the best indication of good experimental rate-distortion performance with a correlation of 0.97 and 0.92 for "lena" and "barbara", respectively. Remarkably, this correlation exists (and is even higher) at low rates where the high resolution assumptions used to develop the coding gain expression are no longer valid. On the other hand, while all of the filters that performed well had reasonably good one-dimensional coding gains, the filters with the best performance generally did *not* maximize that criterion, emphasizing the difficulty in finding a single, reliable criterion for determining the best wavelet basis functions for image coding.

After coding gain, the next best indicator of good experimental performance is the low-band reconstruction error. Our results indicate that the prediction performance improves slightly as the number of levels used to compute $LBRE$ increases. For example, $LBRE$ exhibits a correlation of 0.822, 0.886, and 0.897 for a one, three, and five level decomposition, respectively. Another important observation from Figure 3 is that the low-band reconstruction error predicts actual coding performance more accurately for the orthogonal filters. For instance, when implementing a five level decomposition, the $LBRE$ correlation coefficient is 0.98 for the orthogonal filters, while only 0.89 for the linear phase biorthogonal. Thus, when restricting oneself to orthogonal filters, the $LBRE$ is essentially as good a measure as the coding gain. Conversely, while it is important to have good $LBRE$ for the linear phase biorthogonal filters, it is not sufficient.

Using Table 6 and the scatter plots in Figures 3 and 4, we can see that synthesis regularity follows low-band reconstruction error as the next best predictor of coding performance with a correlation coefficient of 0.75 for both images. It is interesting to note that these

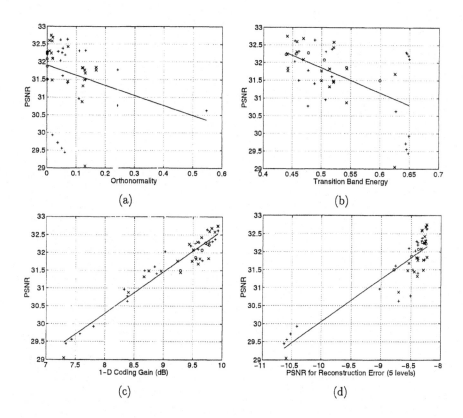

Figure 3. Comparison of actual coding performance with filter bank (a) orthonormality, (b) transition band energy, (c) one-dimensional coding gain, and (d) low-band reconstruction error (5 levels). Orthonormal, odd-length biorthogonal, and even-length biorthogonal filters are labeled by o's, x's, and +'s respectively. Coding performance was measured by the average PSNR for the lena and barbara image over several rates between 0.1 and 1.0 bpp.

relationships seem to confirm the intuition of previous authors that the regularity of the reconstruction wavelets is more important than the analysis regularity and vanishing moments [10], [22]. While the importance of synthesis regularity over analysis regularity is consistent for both the odd and even-length biorthogonal filters, a similar claim is more difficult to make for vanishing moments. The scatter plot of all filter pairs in Figure 4 and the corresponding line-fit seem to indicate that the possession of vanishing moments in the analysis wavelet is more critical than in the synthesis wavelet. However, after a more careful inspection of the scatter plot and a check of the correlation coefficients in Table 6, the opposite appears to be true for the odd-length biorthogonal filters. Again, the experimental results indicate that regularity (synthesis) is the more reliable measure. On the other hand, unlike coding gain, high regularity alone is not sufficient for good performance as there exists filters with both very high regularity and poor coding performance.

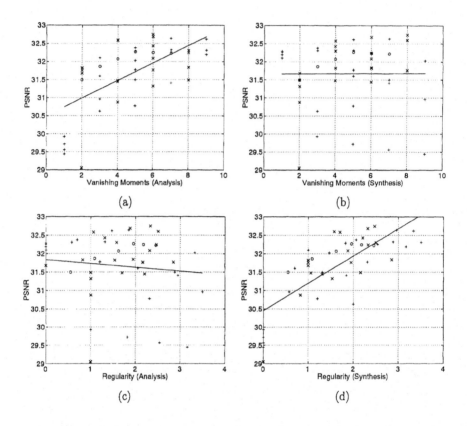

Figure 4. Comparison of actual coding performance with (a) analysis vanishing moments, (b) synthesis vanishing moments, (c) analysis regularity, and (d) synthesis regularity. Orthonormal, odd-length biorthogonal, and even-length biorthogonal filters are labeled by o's, x's, and +'s respectively. Coding performance was measured by the average PSNR for the lena and barbara image over several rates between 0.1 and 1.0 bpp.

For example, consider the factorization of the binomial product filter $P_n(z)$ into the linear phase biorthogonal pair, $H_0(z) = 1$ and $G_0(z) = P_n(z)$. For large n, we can achieve high synthesis regularity (and analysis vanishing moments), even though the resulting coding performance will be quite poor.

From the plots in Figure 2, we can see that with respect to the biorthogonal filters, the largest improvement in PSNR occurs from $n = 4$ to $n = 8$ with an average improvement of approximately 0.5 dB. The general trend appears to be marginally better PSNR performance for longer filters with a break in monotonicity at $n = 10$. This break coincides with the only drop in coding gain for increasing values of n, and is also the only value of n for which the biorthogonal filter coding gain is less than the coding gain for the corresponding Daubechies orthonormal factorization. As a general trend for the orthogonal filters, the rate-distortion performance improves with the increasing order of $P(z)$. This tendency is true for filters with lengths of 4, 6, 8, and 10. However, for filter lengths greater than or

Table 5. Filter Coefficients

6/10	
h_0	g_0
0.78848562	0.61505077
0.04769893	0.13338923
−0.12907777	−0.06723693
	0.00698950
	0.01891423

9/11	
h_0	g_0
0.73666018	0.89950611
0.34560528	0.47680326
−0.05446378	−0.09350470
0.00794811	−0.13670658
0.03968709	−0.00269497
	0.01345671

13/11	
h_0	g_0
0.76724517	0.83284756
0.38326926	0.44810860
−0.06887811	−0.06916271
−0.03347509	−0.10873737
0.04728176	0.00629232
0.00375921	0.01418216
−0.00847282	

17/11	
h_0	g_0
0.82592300	0.75890773
0.42079628	0.41784911
−0.09405920	−0.04036798
−0.07726317	−0.07872200
0.04973290	0.01446750
0.01193457	0.01442628
−0.01699064	
−0.00191429	
0.00190883	

equal to 12 the changes in PSNR become insignificant. This same observation can also be made for the linear phase biorthogonal filters. One notable dissimilarity between the biorthogonal and orthogonal filters is the nature of the frequency responses as n grows large. For example, in Figure 5(a) we note that the Daubechies orthonormal filters begin to approach an ideal brick-wall filter as n increases. In contrast, the frequency responses of the good biorthogonal filters for the same n remain quite asymmetric as evidenced by Figure 5(b). Although the results are not shown here, one additional comment is that overall differences between the rate-distortion curves of the orthogonal filters are most pronounced at lower bit rates and almost nonexistent for rates above 2.0 bpp.

From the empirical results of this section, it is clear that the constraint of exact orthogonality is not necessary for filters derived from the factorization of binomial $P(z)$. In contrast, filter pairs with a center frequency located at $\omega > 0.5\pi$ generally achieved the best rate-distortion performance. This behavior can be partially explained by the impact of aliasing in the subbands. For instance, after one stage of a two-band filter bank decomposition, the filtered image is separated into distinct low and high frequency subbands followed by a factor of two decimation which introduces aliasing into each subband. This aliasing comes in the form of high frequencies being mapped into the low frequency subband and low frequencies being mapped into the high frequency subband. However, the energy distribution in a typical image is significantly higher at lower frequencies, dropping off dramatically with increasing frequency. This skewed energy distribution suggests that the impact of high

Table 6. Correlation coefficients between actual coding performance as measured in PSNR and the various filter attributes discussed in Section 3. Results are tabulated for: (a) all filters, (b) odd-length, linear phase biorthogonal filters, (c) even-length, linear phase biorthogonal filters, and (d) minimum phase orthonormal filters.

Image	ON	TBE	CG	$LBRE$	VAN_A	VAN_S	REG_A	REG_S
Lena	−0.18	−0.50	0.97	0.91	0.58	−0.05	−0.20	0.75
Barb	−0.39	−0.60	0.92	0.85	0.66	0.06	−0.01	0.75
Both	−0.29	−0.56	0.96	0.90	0.63	0.00	−0.11	0.76

(a) All filters

Image	ON	TBE	CG	$LBRE$	VAN_A	VAN_S	REG_A	REG_S
Lena	−0.46	−0.57	0.97	0.88	0.26	0.52	0.29	0.64
Barb	−0.62	−0.73	0.87	0.84	0.47	0.54	0.44	0.76
Both	−0.54	−0.66	0.94	0.87	0.37	0.53	0.37	0.71

(b) Odd length, linear phase biorthogonal filters

Image	ON	TBE	CG	$LBRE$	VAN_A	VAN_S	REG_A	REG_S
Lena	−0.01	−0.42	0.98	0.93	0.80	−0.41	−0.47	0.85
Barb	−0.26	−0.49	0.94	0.87	0.82	−0.28	−0.29	0.79
Both	−0.13	−0.46	0.98	0.91	0.82	−0.36	−0.39	0.83

(c) Even length, linear phase biorthogonal filters

Image	TBE	CG	$LBRE$	VAN	REG
Lena	−0.91	0.95	0.96	0.80	0.89
Barb	−0.60	0.98	0.99	0.91	0.97
Both	−0.96	0.99	0.98	0.88	0.95

(d) Minimum phase, orthonormal filters

frequency aliasing into the low frequency subband is considerably less substantial than the aliasing of low frequency information into the high frequency subband. Consequently, the results of aliasing in the high frequency subband may be quite large when compared to the actual signal energy, increasing the number of nonzero coefficients to be quantized and subsequently reducing the number of zero-trees. Shifting the transition band beyond 0.5π alleviates aliasing of low frequency components into the high frequency subband at the expense of increased aliasing into the low frequency subband. Experimentally, for image data this type of behavior is more desirable. One important note is that the filter order cannot be arbitrarily increased so as to achieve extremely sharp frequency responses for the

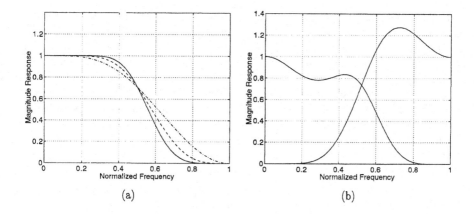

Figure 5. Comparison of the frequency responses for the (a) length 4 (dash-dot), 8 (dash), and 12 (solid) Daubechies lowpass analysis/synthesis filters, and the (b) linear phase biorthogonal 16/12 lowpass and highpass analysis filters.

elimination of aliasing due to the extremely high ringing that would occur at even modest compression ratios.

4.2.3. Complexity and Perceptual Issues

In addition to quantitative PSNR comparisons, the reconstructed images were evaluated to assess the visibility of coding errors. For our experiments, ringing generally constituted the most disturbing visual artifact. As a general rule, images coded using the linear phase biorthogonal filters exhibited significantly less ringing for a given order of $P(z)$ in addition to having better PSNR. We can explain this behavior by recalling that for a biorthogonal factorization the filters in the analysis and synthesis section are allowed to differ while they must remain identical for orthonormal filter banks. In image coding, the ringing can generally be attributed to a sharp transition band in the lowpass synthesis filter. Thus, for linear phase biorthogonal filters the lowpass filter in the analysis section can have a sharper cutoff for better coding gain while allowing a more gentle transition band in the synthesis section for less ringing.

Often, the better rate-distortion performance of the longer linear phase filters was overshadowed by substantial ringing, and the short 5/3 and 2/6 filters produced the most pleasing images, especially at bit rates below 0.5 bpp, while for the orthonormal filters, significant ringing was visible for filters as short as six. Experimental results demonstrate that, surprisingly, even the simple 5/3 and 2/6 filter pairs provided good rate-vs.-PSNR performance. For the embedded zero-tree quantization strategy that we employed, the 2/6 pair outperformed all of the orthogonal filters tested at rates below 1.0 bpp for the "lena" image. Furthermore, because of the simplicity of their coefficients, the implementation of both the 5/3 and 2/6 filters requires no multipliers and can be achieved with simple integer additions and shift operations. On the more complex "barbara" image, the 2/6 filter performed better than orthonormal filters with lengths of 4 and 6, but was outmatched by the longer filters. In general, our experiments suggest that the longer biorthogonal and orthonormal filters provide the greatest gains for images with large amounts of texture or edges. For instance, many of the larger filters were able to resolve the fine lines and texture on the well-known stripes in the "barbara" image, whereas the shorter filters produced somewhat disturbing stair-stepping along edges. On the other hand, the shorter filters provided the most competitive PSNR results for the simpler "lena" image and typically produced less ringing.

5. Conclusions

From the set of all possible factorizations of binomial product filters for $n \leq 14$, candidate filters were selected based on the filter attributes described in Section 3 with the goal of identifying filters with good coding performance. From this pool, the best filters for each n were determined by experimental rate-distortion performance. Table 5 contains the coefficients of the best filters (normalized to $\sqrt{2}$) for each n that are less well known or not currently available in the literature. For example, while the 6/10 and 13/11 filters have also been presented by Villasenor et al. in [22], they do not enjoy the wide spread popularity of other filter pairs like the 9/7. Similarly, the coefficients of the 9/11 filter pair have been tabulated by Daubechies [7], p. 279, although the pair in that case was not specifically recommended for use with image coding. Additionally, the order of the filters were inverted so that the coefficients were presented as an 11/9 pair. Finally, we note that the filter pair with the best performance, the 17/11 filter pair, is to our knowledge not currently available elsewhere in the literature, and is presented for the first time in [24] and Table 5.

Coding experiments conducted on image data indicate that a number of properties seem to be important in the design of linear phase biorthogonal filters for data compression, including degree of orthonormality, stopband attenuation, coding gain, low-band reconstruction error, synthesis regularity, and analysis vanishing moments. By far, the one-dimensional coding gain is the best predictor of experimental rate-distortion performance, followed by the low-band reconstruction error, and synthesis regularity. However, no single attribute, including coding gain, uniquely determines the best filter for actual use. Moreover, the importance of each filter attribute varies depending on whether the filters are orthonormal, odd-length biorthogonal, or even-length biorthogonal.

From the experimental results, the superiority of linear phase biorthogonal filters over corresponding orthonormal filters suggests that while some orthonormality is important, it is not necessary to direct all degrees of design freedom towards the goal of absolute orthonormality. This assertion is consistent with our previous observation that for an octave-band decomposition, the coding gain of some linear phase biorthogonal filters exceeds the maximum coding gain achievable by any orthonormal filter [24].

Acknowledgments

Most of the research described in this paper was performed at the Jet Propulsion Laboratory, California Institute of Technology, under contract with NASA. This material is also based upon work supported in part under a National Science Foundation Graduate Fellowship and a University of California MICRO grant with matching support from Hughes Aircraft, Signal Technology Inc., and Xerox Corporation. The authors thank Charles D. Creusere at the Naval Air Warfare Center in China Lake, CA, for providing them with the wavelet quantization routines, and Ali Akansu, Cormac Herley, Michael Unser, P. P. Vaidyanathan, and Victor Wickenhauser for many helpful discussions during the course of this research.

Notes

1. Throughout this paper we use the notation n_1/n_2 to denote linear phase biorthogonal filters for which the lengths of $H_0(z)$ and $G_0(z)$ are n_1 and n_2, respectively. It is important to mention, however, that this nomenclature does not uniquely specify the factorization in general.

References

1. P. P. Vaidyanathan, *Multirate Systems and Filter Banks*, Englewood Cliffs, NJ: Prentice Hall, 1993.

2. M. Vetterli and C. Herley, "Wavelets and filter banks: theory and design," *IEEE Trans. on Signal Processing*, vol. 40, 1992, pp. 2207–2232.

3. J. D. Johnston, "A filter family designed for use in quadrature mirror filter banks," in *Proc. of the Int. Conf. on Acoust. Speech and Sig. Proc.*, 1980, pp. 291–294.

4. F. Grenez, "Chebyshev design of filters for subband coders," *IEEE Trans. Acoust., Speech, and Signal Processing*, vol. 36, 1988, pp. 182–185.

5. M. J. T. Smith and T. P. Barnwell, "Exact reconstruction techniques for tree-structured subband coders," *IEEE Trans. Acoust., Speech, and Signal Processing*, vol. 34, 1986, pp. 434–441.

6. J. W. Woods and S. D. O'Neil, "Subband coding of images," *IEEE Trans. Acoust., Speech, and Signal Processing*, vol. 34, 1986, pp. 1278–1288.

7. I. Daubechies, *Ten Lectures on Wavelets*, Philadelphia, PA: SIAM, 1992.

8. A. Gersho and R. M. Gray, *Vector Quantization and Signal Compression*, Boston: Kluwer Academic Publishers, 1992.

9. J. Katto and Y. Yasuda, "Performance evaluation of subband coding and optimization of its filter coefficients," in *Proc. of the SPIE Symposium on Visual Comm. and Image Proc.*, vol. 1605, 1991, pp. 95–106.

10. M. Antonini, M. Barlaud, P. Mathieu, and I. Daubechies, "Image coding using the wavelet transform," *IEEE Trans. on Image Processing*, vol. 1, 1992, pp. 205–220.

11. I. Daubechies, "Orthonormal bases of compactly supported wavelets," *Communications on Pure and Applied Mathematics*, vol. XLI, 1988, pp. 909–996.

12. D. L. Gall and A. Tabatabai, "Subband coding of digital images using symmetric short kernal filters and arithmetic coding techniques," in *Proc. of the Int. Conf. on Acoust. Speech and Sig. Proc.*, 1988, pp. 761–764.

13. R. Ansari and D. L. Gall, "Advanced television coding using exact reconstruction filter banks," in J. W. Woods (ed.), *Subband Image Coding*, Boston, MA: Kluwer Academic Publishers, 1991, pp. 273–318.

14. E. Majani and M. Lightstone, "Wavelet compression of seismic data," *Int. Conf. on Wavelets, Theory, Algorithms, and Applications*, Taormina, Italy, 1993, pp. 14–20.

15. M. Lightstone and E. Majani, "Low bit-rate design considerations for wavelet-based image coding," in *Proc. of the SPIE Symposium on Visual Comm. and Image Proc.*, vol. 2308, Chicago, IL, 1994, pp. 501–512.

16. E. Majani and M. Lightstone, "Biorthogonal wavelets for data compression," in *Proceedings of the Data Compression Conference*, Snowbird, UT, 1994, p. 462.

17. J. N. Bradley, C. M. Brislawn, and T. Hopper, "The FBI wavelet/scalar quantization for gray-scale fingerprint image compression," *Los Alamos Technical Report*, vol. LA-UR-93-1659.

18. A. V. Oppenheim and R. W. Schafer, *Discrete-time signal processing*, Englewood Cliffs, NJ: Prentice-Hall, 1989.

19. A. K. Soman and P. P. Vaidyanathan, "Coding gain in paraunitary analysis/synthesis systems," *IEEE Trans. on Signal Processing*, vol. 41, 1993, pp. 1824–1835.

20. I. Djokovic and P. P. Vaidyanathan, "Statistical wavelet and filter bank optimization," in *Conference Record of The Twenty-Seventh Asilomar Conference on Signals, Systems and Computers*, vol. 2, 1993, pp. 911–915.

21. N. S. Jayant and P. Noll, *Digital coding of waveforms*, Englewood Cliff, New Jersey: Prentice-Hall, 1984.

22. J. D. Villasenor, B. Belzer, and J. Liao, "Filter evaluation and selection in wavelet image compression," in *Proceedings of the Data Compression Conference*, Snowbird, UT, 1994, pp. 351–360.

23. O. Rioul, "Simple regularity criteria for subdivision schemes," *SIAM J. Math. Anal.*, vol. 23, 1992, pp. 1544–1576.

24. E. Majani, "Biorthogonal wavelets for image compression," in *Proc. of the SPIE Symposium on Visual Comm. and Image Proc.*, vol. 2308, Part 1, Chicago, IL, 1994, pp. 478–488.

25. R. M. de Queiroz and H. S. Malvar, "On the asymptotic performance of hierarchical transforms," *IEEE Trans. on Signal Processing*, vol. 40, 1992, pp. 2620–2622.

26. J. M. Shapiro, "An embedded wavelet hierarchical image coder," in *Proc. of the Int. Conf. on Acoust. Speech and Sig. Proc.*, 1992, p. 1992.

27. J. M. Shapiro, "Embedded image coding using zerotrees of wavelet coefficients," *IEEE Trans. on Signal Processing*, vol. 41, 1993, pp. 3445–3462.

28. C. Herley and M. Vetterli, "Orthogonal time-varying filter banks and wavelet packets," *IEEE Trans. on Signal Processing*, vol. 42, 1994, pp. 2650–2663.

29. M. J. T. Smith and S. L. Eddins, "Analysis/synthesis techniques for subband image coding," *IEEE Trans. Acoust., Speech, and Signal Processing*, vol. 38, 1990, pp. 1446–1456.

30. G. Karlsson and M. Vetterli, "Extension of finite length signals for subband image coding," *Signal Processing*, vol. 17, 1989, pp. 161–168.

Multidimensional Systems and Signal Processing, 8, 129–150 (1997)

Multiresolution Vector Quantization for Video Coding

G. CALVAGNO   calvagno@dei.unipd.it
Dipartimento di Elettronica e Informatica, Università di Padova, 35131 Padova, Italy

R. RINALDO   rinaldo@dei.unipd.it
Dipartimento di Elettronica e Informatica, Università di Padova, 35131 Padova, Italy

Abstract. In this work, we propose a coding technique that is based on the generalized block prediction of the multiresolution subband decomposition of motion compensated difference image frames. A segmentation mask is used to distinguish between the regions where motion compensation was effective and those regions where the motion model did not succeed. The difference image is decomposed into a multiresolution pyramid of subbands where the highest resolution subbands are divided into two regions, based on the information given by the segmentation mask. Only the coefficients of the regions corresponding to the motion model failure are considered in the highest resolution subbands. The remaining coefficients are coded using a multiresolution vector quantization scheme that exploits inter-band non-linear redundancy. In particular, blocks in one subimage are predicted from blocks of the adjacent lower resolution subimage with the same orientation. This set of blocks plays the role of a codebook built from coefficients inside the subband decomposition itself. Whenever the inter-band prediction does not give satisfactory results with respect to a target quality, the block coefficients are quantized using a lattice vector quantizer for a Laplacian source.

Key Words: video coding, vector quantization, low bit rate, subband coding

1. Introduction

Very low bit rate video coding is receiving very much attention due to its wide range of possible applications [1].

Model-based coding techniques have the most promising capability in achieving satisfactory image quality at very high compression ratios due to the possibility of describing image sequences using a limited set of parameters [2]. Nevertheless, these coding schemes fail every time the video signal to be coded does not fit the model. Moreover, allowing a sufficiently general model for the image sequence dramatically reduces the coder performances, since the amount of data necessary to describe complex models increases rapidly.

Waveform-based coding techniques are therefore still good candidates for low bitrate video transmission and storage, especially when the image sequence source is not constrained to fit any particular model. Moreover, in a more advanced setting, where different coding algorithms will be integrated to form a "universal" coding system, which is capable of switching between the various techniques according to the input signal characteristics, waveform-based coders will certainly play a very important role.

To achieve the goal of very low bitrate transmission, a waveform based coding system should conveniently take advantage of both the spatial and temporal redundancy present in typical video signals [1]. In this paper we propose a video coder that is based on a hybrid temporal and spatial compression scheme.

The temporal redundancy of the image sequence to be coded is reduced using motion-compensated Differential Pulse Code Modulation (DPCM) to generate a sequence of inter-frame difference images.

The residual spatial redundancy that is still present in the difference signal is then further reduced by means of a subband coding technique that uses a multiresolution decomposition of the inter-frame images [3]. In this algorithm, the similarity among the subbands of different levels in a pyramid subband decomposition of the inter-frame image is exploited to provide efficient data compression. Thus, the technique takes advantage of the non-linear redundancy among multiresolution subimages. This is done by partitioning the pyramid representation into square blocks and trying to predict blocks in one subimage from blocks of the adjacent lower resolution subimage with the same orientation. The lower resolution subimage plays therefore the role of a codebook built from coefficients inside the subband decomposition itself. Whenever this inter-band prediction does not give satisfactory results with respect to a target quality, the block coefficients are quantized using a lattice vector quantizer.

To fully exploit the temporal and spatial redundancy, the information provided by the motion compensation procedure is used to increase the efficiency of the multiresolution coding algorithm. Namely, the block matching procedure, which performs the motion compensation, produces a segmentation mask that discriminates between blocks for which the motion compensation is successful and blocks where the motion compensation is not satisfactory. On the basis of this mask, only the coefficients of blocks in the subimages at the two lowest levels of the multiresolution decomposition, corresponding to the region where motion compensation fails, are actually coded while the others are set to zero.

The proposed coding system has been tested using typical video conference image sequences with frames of size 352×288 pixels, at a rate of 30 frame/s (CIF format). The simulation results show that it can be used for video transmission at rate ranging from 48 to 128 kbit/s with good visual quality.

The organization of the paper is as follows. Section 2 is a short review of subband decomposition principles. Section 3 contains a short description of the multiresolution coding algorithm that is used to compress the inter-frame difference images. Section 4 describe the entire hybrid temporal and spatial compression scheme in detail and Section 5 reports the obtained coding results. The conclusions are drawn in Section 6.

2. Subband Decomposition Principles

In this section we briefly review subband coding principles. We refer the interested reader to [4] for a more complete treatment.

Subband decomposition for images has first been introduced by Woods and O'Neil [5]. A popular scheme for two-dimensional subband decomposition is based on separable filters, as shown in Figure 1. Subband x^{ij}, $i, j = 0, 1$, is obtained by first filtering the rows of the input image with $H_i(z)$ and subsampling by a factor two, followed by filtering the columns with $H_j(z)$ and subsampling. In the scheme of Figure 1, $H_0(z)$ is a low-pass filter, while $H_1(z)$ has high-pass characteristics: thus, each subband x^{ij} has one fourth of the input image coefficients and is relative to details that pertain to different frequency regions of the input image spectrum.

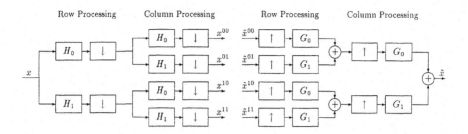

Figure 1. Separable 2D scheme for subband decomposition.

Assuming no coding error, the reconstruction is performed at the receiver by upsampling the rows of x^{ij}, and filtering with $G_i(z)$, followed by upsampling the columns and filtering with $G_j(z)$. The outputs of the four upsample/filter sections are summed together to give the reconstructed image $\tilde{x}$. It is possible to show that in the absence of coding errors, perfect reconstruction of the input image x can be obtained by appropriate design of the analysis filters $H_i(z)$ and synthesis filters $G_i(z)$ [4], [6], [7].

Paraunitary systems [6] are perfect reconstruction systems where the filter bank can be obtained by designing a single filter $H_0(z)$. In fact, one possible way to achieve perfect reconstruction is to choose the analysis/synthesis filter coefficients such that

$$
\begin{aligned}
g_0(n) &= h_0(-n) \\
g_1(n) &= h_1(-n) \\
h_1(n) &= (-1)^{1-n} h_0(1-n),
\end{aligned}
\tag{1}
$$

where $h_0(n)$ is a filter whose z-transform $H_0(z)$ satisfies

$$
H_0(z)H_0(z^{-1}) + H_0(-z)H_0(-z^{-1}) = 2.
\tag{2}
$$

Note that relation (2) implies that the filter $h_0(n)$ and its even translations form an orthonormal family, while relations (1) imply that the two filters $h_0(n)$ and $h_1(n)$ are orthogonal.

In [8], [9] it is shown that orthonormal bases of wavelets correspond to a subband coding scheme with orthogonal filters satisfying equations (1) and (2). Considering the case of FIR filters, it is possible to show [6] that conditions (1) and (2) can be met exactly only for even-length filters. Furthermore, only trivial solutions with linear phase are possible. Approximate solutions with odd length linear phase FIR filters can be found in [9], [10]. A scheme with linear phase FIR analysis and synthesis filters has been proposed in [11], and a design method with several examples can be found in [12]. A consequence of the linear phase constraint is that relation (2) can only be satisfied approximately. More general solutions for exact perfect reconstruction even-length, odd-length FIR linear phase filters are given by biorthogonal systems [6], [7], where the orthogonality conditions (1) and (2) are relaxed.

The basic separable two-dimensional scheme of Figure 1 can be used to obtain finer decompositions of the input spectrum. In a pyramidal scheme [9], the decomposition of Figure 1 is iterated to obtain a multiresolution decomposition of $x(m, n)$. Denoting with x_0^{00}

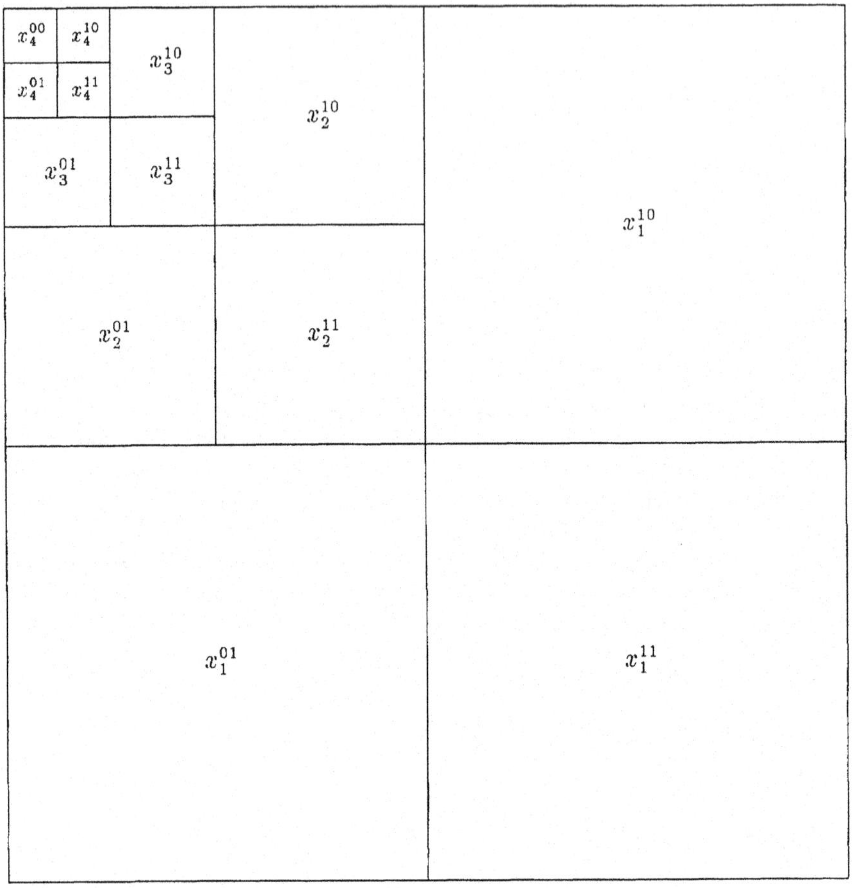

Figure 2. Organization of the subimages in a pyramid subband dacomposition.

the input image $x(m, n)$, at each decomposition level l the image x_{l-1}^{00} is decomposed into the four subimages $x_l^{00}, x_l^{01}, x_l^{10}, x_l^{11}$, for $l = 1, \ldots, l_M$. The result of such a decomposition is a set of subimages which are localized in scale, orientation and space. Figure 2 shows the organization of the subimages in a 4 level decomposition.

In the coder, we apply a hybrid vector quantization scheme to a multiresolution subband decomposition of the input image that makes explicit use of the similarity inherent in the local structure of block coefficients in different subbands, i.e., of the inter-band redundancy/non-linear dependence of subbands. The next section presents the basic principles of our scheme.

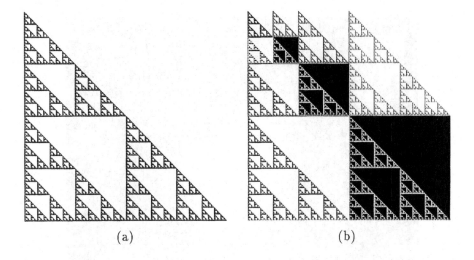

Figure 3. (a) The "Sierpinsky Triangle"; (b) Wavelet Transform of the "Sierpinsky Triangle."

3. Multiresolution Vector Quantization

An interesting property of the pyramid subband decomposition is that subimages with the same orientation show similar *local* behavior around the edges of the image.

The redundancy of the wavelet transform has been recognized by several authors [13], [14], [15], [16]. A striking example is given by Fig. 3, where the Sierpinsky Triangle [17] and its 3 level decomposition, computed with the 4 tap Daubechies' wavelet [8], are shown. For visual convenience, the values of the coefficients in each subband are appropriately scaled to cover the full gray scale. It is evident from the figure that the edges in the original image originate high energy coefficients in each subband: moreover, the actual shape of the coefficient surface is very similar from one scale to the other. Although the intrinsic redundancy of the wavelet decomposition for fractal images is apparent from Fig. 3, it can be noted also for generic non-fractal images and different filter banks. Fig. 4 shows frame number 1 of the image sequence "Miss America," while Fig. 5 shows the three lowest resolution level subbands of its 4 level multiresolution decomposition corresponding to a separable filter bank based on the nearly orthogonal, almost perfect reconstruction 9-tap linear phase filters of [10]. The original 8-bit greyscale video sequence is in CIF format (352×288 pixels), at a rate of 30 frame/s. From Fig. 5, it is evident that it is possible to find blocks with similar characteristics inside the subregions located in adjacent subbands.

Based on these considerations, we designed a hybrid vector quantization coding scheme for still images [3] in which *range* blocks of subband coefficients are predicted from *domain* blocks of the same dimensions located in the subband with the same orientation and at a lower resolution. Thus, the pool of domain blocks acts as a *codebook* for the range block, as in vector quantization, with the important difference that the codebook is built from blocks

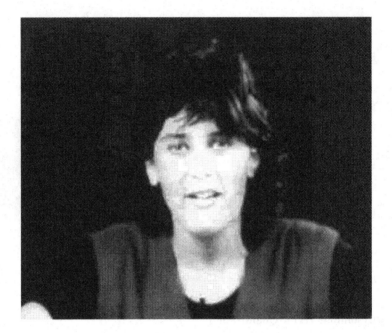

Figure 4. Frame 1 of "Miss America": Original image.

inside the subband decomposition itself. To provide initial conditions to the prediction scheme, the coefficients of the subimages at the lowest level of the pyramid decomposition, namely $x_{l_M}^{11}$, $x_{l_M}^{01}$, $x_{l_M}^{10}$ are modeled as independent Laplacian random variables [5] and quantized using an adaptive version of the pyramid vector quantizer (PVQ) described in [18], [19]. PVQ is also used in higher resolution subbands whenever a range block can not be accurately predicted from the domain block codebook.

In this paper, we propose an extension of the algorithm of [3] that is suitable for image sequence coding. In particular, in this section we review the technique for two-dimensional signals and describe it for still images in CIF format. Its extension for video signals will be considered in the next section. We start by briefly reviewing pyramid vector quantization and the adaptive version we use for our coder [18], [19], [20]. We then describe the hybrid coder.

3.1. Pyramid Vector Quantization

Geometric quantizers [18], [21] exploit the geometric properties of memoryless sources of random variables. Let $f_X(a)$ be the one-dimensional probability density function of a random variable X. Then any L-dimensional vector $\mathbf{x}$ of independent and identically

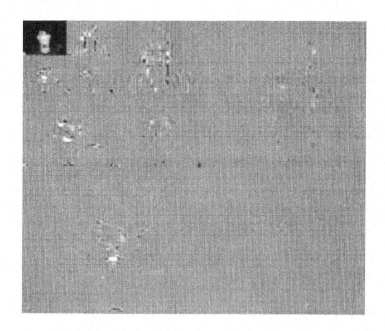

Figure 5. Three lowest resolution level subbands of frame 1 of "Miss America."

distributed (iid) variables drawn according to $f_X(a)$ has a joint probability density function

$$f_{\mathbf{x}}(\mathbf{a}) = \prod_{i=1}^{L} f_X(a_i). \tag{3}$$

Taking the logarithm of (3), we can write the following equation between functions of the random vector $\mathbf{x} = (X_1, \ldots, X_L)^T$:

$$\frac{1}{L} \log_2 f_{\mathbf{x}}(\mathbf{x}) = \frac{1}{L} \sum_{i=1}^{L} \log_2 f_X(X_i). \tag{4}$$

The weak law of large numbers assures that the arithmetic mean in (4) converges in probability to the expectation of $\log_2 f_X(X)$ as $L \to \infty$, i.e.,

$$\frac{1}{L} \log_2 f_{\mathbf{x}}(\mathbf{x}) \xrightarrow{P} \int f_X(a) \log_2 f_X(a) da \triangleq -h(X). \tag{5}$$

Here, $h(X)$ denotes the differential entropy of the scalar random variable X. Relation (5) can be rewritten equivalently by stating that the *typical set* [22]

$$S_\epsilon = \{\mathbf{a} : 2^{-L(h(X)-\epsilon)} < f_{\mathbf{x}}(\mathbf{a}) < 2^{-L(h(X)+\epsilon)}\} \tag{6}$$

has a probability

$$P[S_\epsilon] = \int_{S_\epsilon} f_\mathbf{X}(\mathbf{a})d\mathbf{a}$$

that can be made arbitrarily close to 1 as $L \to \infty$, for any ϵ. Thus, relation (5) implies that, for large L, vectors of iid random variables drawn according to $f_X(a)$ concentrate with high probability in a subset of $\mathbb{R}^L$ where

$$f_\mathbf{X}(\mathbf{a}) \simeq 2^{-Lh(X)}. \tag{7}$$

Therefore, for random vectors of iid variables, the reproduction points of a vector quantizer should be uniformly located on the set S_ϵ, which can be determined once the probability density $f_X(a)$ is known.

Based on these results, Fischer [18] designed a *pyramid* vector quantizer for L-dimensional vectors of iid laplacian random variables. In this case the joint probability density is given by

$$f_\mathbf{X}(\mathbf{a}) = \left(\frac{\lambda}{2}\right)^L \exp(-\lambda \sum_{i=1}^{L} |a_i|) \tag{8}$$

and the entropy per degree of freedom is

$$h(X) = \log_2 \left(\frac{2e}{\lambda}\right). \tag{9}$$

By substituting (8) and (9) in (7) it turns out that the relevant volume for quantization is located around the L-dimensional hyper-pyramid defined by

$$\sum_{i=1}^{L} |a_i| = \frac{L}{\lambda}. \tag{10}$$

In [23] a variation of the PVQ has been presented to code vectors of subband coefficients which have approximately a laplacian distribution [5]. The PVQ codebook vectors are determined by the intersection of a cubic lattice with a scaled L-dimensional pyramid. The pyramid is selected to be close to the nominal one, defined by (10), on the basis of the value

$$||\mathbf{x}||_1 = \sum_{i=1}^{L} |X_i| \tag{11}$$

computed from the actual components of each input vector $\mathbf{x}$. A product gain-shape code is used, in which $||\mathbf{x}||_1$ is coded using a Lloyd-Max gaussian quantizer to index the actual pyramid, and the remaining Lr bits are used to identify the lattice point on the pyramid. At the decoder, we need the nominal variance $\sigma^2 = 2/\lambda^2$, a code for $||\mathbf{x}||_1$ and rL bits per vector of input coefficients [18], [23]. It can be shown that the mean squared error distortion obtained with such PVQ is

$$D_{PVQ}(r) \simeq \frac{e^2}{6}\sigma^2 2^{-2r} \tag{12}$$

and that the asymptotic performance of the PVQ is the same of uniform quantization followed by entropy coding [24].

An adaptive version of the PVQ is obtained by imposing a target distortion and determining the rate r_k for vector $\mathbf{x}_k$ on the basis of the asymptotic rate-distortion performance of the PVQ given by (12), and of the sample variance of the input coefficients

$$\sigma_k^2 = \frac{2}{\lambda_k^2} = \sum_i X_{k,i}^2. \tag{13}$$

In our implementation, the rate r_k is uniformly quantized with step 0.25 bit. The vector of coefficients is fed into the PVQ using a nominal value for the variance that is recalculated from $D_{PVQ}(r)$ and the quantized values r_k. We use a fixed number N_r of bits for $||\mathbf{x}||_1$. Thus, besides the $N_r + r_k L$ bits for each input vector, at the decoder we need side information about the quantized values r_k.

3.2. The hybrid coder

We now describe the organization of the hybrid coder. For 352×288 CIF images, we compute $l_M = 4$ levels of the multiresolution subband decomposition. Each 22×18 subband $x_4^{i,j}$ is divided into six 11×6 blocks. The coefficients of these blocks are used to form the input vectors that are coded using adaptive PVQ. The imposed distortion is set to a value T that determines the target quality of the reconstructed image. Using 66 coefficient vectors guarantees performance close to the asymptotic values. Baseband x_4^{00} coefficients have a uniform distribution and some residual correlation. In order to have coefficients with a Laplacian distribution, they are first DCT transformed before coding using PVQ.

The coded subimages at level $l_M = 4$ of the multiresolution decomposition are used to form the codebook for subband coefficient blocks inside the higher resolution subimages at level $l = 3$. Specifically, subband x_3^{01} (or x_3^{10}, x_3^{11}) is subdivided into non-overlapping 4×4 range blocks. A domain block d_h of the same dimension is searched in x_4^{01} (or x_4^{10}, x_4^{11}) to minimize the mean square error (MSE) distance

$$D(b_k, d_h) = \sum_l \sum_m (b_k(l, m) - \tau_k[d_h](l, m))^2 \tag{14}$$

between the range block b_k and an appropriately transformed domain block $\tau_k[d_h]$. Here, l, m, indicate the coefficient position inside the block. The transformation τ_k has the form

$$\tau_k[d_h] = \alpha_k(\mathcal{I}_k[d_h]), \tag{15}$$

where α_k is a multiplicative scaling factor and $\mathcal{I}_k$ is one of the possible four rotations of the domain block coefficients, chosen to minimize (14). The transformations are similar to those proposed for fractal block coders [25], [26]. The code for each range block is given by the address of the domain block minimizing distance (14), in addition to the parameters defining the transformation τ_k, i.e., the value of the scaling parameter α_k (uniformly quantized and entropy coded) and the index of the rotation. For a given range block at relative position (j, k) in one subimage, the block matching procedure is based on the domain blocks belonging to a subregion centered at the corresponding relative position $(j/2, k/2)$ in the adjacent lower resolution subband with the same orientation. We use 3 bits per direction to specify the domain block address.

It is of course possible that no domain block provides a good match for a particular range block, i.e., that the MSE distance between the range block and any transformed domain block is greater than the threshold T. In case no good match is found, the L coefficients inside the block are modelled as i.i.d. Laplacian random variables, and coded via adaptive PVQ with an imposed distortion set to T. The value of T plays the role of a qualitative parameter for the coding procedure, where smaller values of T are used to obtain a better quality at the expense of a higher bit rate. The parameter T can be adjusted to meet a target compression or reconstruction quality.

To avoid error propagation, the *quantized* images at level 3 are used as the domain block pool for subbands at level 2. The prediction procedure described above is applied to each range block in subbands at level 3. The dimensions of the range blocks considered for prediction vary depending on the resolution. Specifically, at level 2, we start considering 8×8 range blocks that are predicted from domain blocks of the same dimension. If no satisfactory match is found, the 8×8 block is split into 4 smaller blocks, and the search is repeated until a good match is found or the block is PVQ coded. The same scheme and splitting procedure is applied to the range blocks in x_1^{ij}, where we also consider 16×16 range blocks. The prediction is started at the largest allowed dimension and each block is split into four smaller subblocks every time the prediction quality is not satisfactory. If prediction is still not acceptable at the smallest allowed 4×4 block dimension, the block is coded using PVQ.

Additional bits per block are needed to specify the block type in order to correctly decode the stream of bits relative to each block. For the subbands at level 3, one bit per block is sufficient to distinguish between the predicted and PVQ coded blocks. Each 8×8 block at level two can be split into four 4×4 blocks or predicted as it is. One bit is sufficient to distinguish between these two possibilities. Only in the case the block is split, one additional bit for each of the resulting 4×4 blocks is needed to distinguish between the prediction and PVQ alternatives. Thus, additional bits are used only when necessary. The same strategy is applied to classify blocks of subimages at level one of the multiresolution decomposition. Starting from the largest allowed block dimension, one bit specifies if the block is predicted as it is or it is split. In the latter case, the four originated blocks are recursively classified with one bit indicating prediction or further splitting for the intermediate block sizes, and prediction or PVQ for the smallest block size 4×4.

The proposed scheme appears to be remarkably simple and effective. Moreover, the causality of the coding procedure from low to high resolution subimages allows for progressive transmission of the coded image. This is important in many applications like image retrieval and multimedia applications.

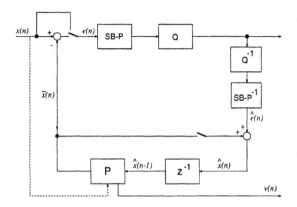

Figure 6. The proposed scheme for video coding (SB-P=subband pyramid).

4.   The Video Coding Scheme

Successive frames of typical video sequences are highly correlated, due to the nature of common scenes, where only a small portion of the picture really changes from frame to frame. It has been recognized [1] that an efficient and effective way to reduce this kind of temporal redundancy is to utilize motion-compensated DPCM to generate inter-frame difference images. The residual redundancy can be further reduced by coding the obtained difference pictures.

The video coding technique proposed in this paper falls into this category of algorithms, where the information given by the motion vectors is also exploited to efficiently code the difference images.

Consider the scheme of Fig. 6. It comprises a DPCM loop to remove the temporal redundancy between image sequence frames. Motion compensation is based on block matching, with 16×16 blocks and one pixel accuracy. Five bits per direction are used for the motion vectors. The search for a matching block in the previously decoded frame is stopped as soon as the MSE is lower than a fixed threshold. In this way, blocks in the static background are classified as not moving, and the influence of noise in the block matching procedure is reduced.

In the present implementation of the coder, the threshold value is set to $T_{MSE} = 20$. This corresponds to a prediction of the block via the motion vector with a Peak Signal to Noise Ratio (PSNR)

$$\text{PSNR} = 10 \log_{10} \frac{255^2}{\text{MSE}} \tag{16}$$

greater than 35 dB, which is an acceptable quality for video conference sequences. In any case, the residual image will be coded as explained below.

Each motion compensated block is coded with the corresponding motion vector, with one additional bit specifying if motion compensation from the previous frame was below the threshold. Thus, the motion vector field is associated with a segmentation mask that permits to identify regions where compensation was effective.

The difference between the current frame and its motion compensated prediction is then subband coded using a multiresolution decomposition, as explained in section 2. The higher frequency subband regions corresponding to image blocks that are motion compensated with a residual error energy lower than the threshold are set to zero (*temporal prediction*). This entails propagating nearly-zero regions in the difference image to its subband decomposition, in order not to code the higher subband coefficients corresponding to image blocks that are already and satisfactorily predicted by motion compensation.

Specifically, difference image coding is performed on the coefficients of a 4 level multiresolution subband decomposition that is computed using the 9-tap zero-phase orthogonal and perfect reconstruction filter bank of [10]. The image is symmetrically extended before filtering to avoid border effects [27]. Temporal prediction consists in setting to zero the blocks of the subband images x_1^{ij} and x_2^{ij} corresponding to the nearly-zero blocks in the motion-compensated image. A 16×16 block at pixel location h, k in the original image corresponds therefore to an 8×8 block at relative position $h/2, k/2$ in x_1^{ij}, and to a 4×4 block at relative position $h/4, k/4$ in x_2^{ij}.

Subbands x_4^{ij}, x_3^{ij} and the remaining blocks that were not set to zero by temporal prediction in subbands x_2^{ij} and x_1^{ij}, are coded with the two-dimensional multiresolution vector quantization (MVQ) algorithm described in Section 3. Note that information to distinguish between difference image blocks that were temporally predicted and those that have to be coded with the MVQ algorithm is provided by the segmentation mask associated with the motion field. An important feature of the algorithm is that blocking effects in the reconstructed frames are reduced because all the coefficients in the residual subimages x_4^{ij} and x_3^{ij} are coded, independently on the MSE accuracy of block matching. This reduces the visual effect of motion compensation with an incorrect motion field.

The differences between motion vectors of adjacent blocks and the segmentation mask are coded using a dictionary based coder [28].

5. Experimental Results

In this section, we present some coding results of the technique described above. We consider first the case of still images.

Figure 7 shows the original frame 5 of the image sequence "Claire." The original image is in CIF format (352×288 pixels). In Figure 8, we plot the PSNR for the reconstructed image as a function of the bit rate, in bits per pixel (bpp).

In the same plot, the proposed technique is compared with the coding standard JPEG [29]. Notice that the proposed MVQ coder performs better than JPEG over the entire range of bit rates, with an improvement in PSNR of about 2 dB, almost independent of the bit rate.

Figure 9a shows the reconstructed frame 5 of "Claire," coded at 0.19 bpp. In Figure 9b, we show the same frame coded with JPEG, again at 0.19 bpp. Despite some ringing artifacts, it can be noticed from Figure 9a that no blocking effect is visible in the reconstructed image,

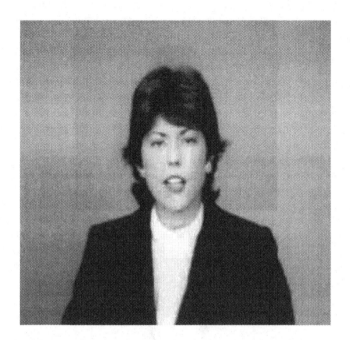

Figure 7. Frame 5 of "Claire": Original image.

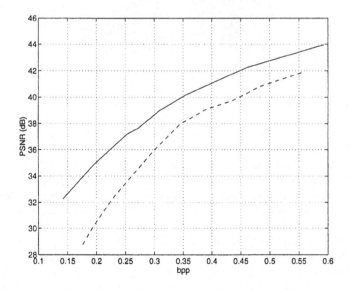

Figure 8. PSNR vs. bpp using MVQ coding (solid line) and JPEG (dashed line).

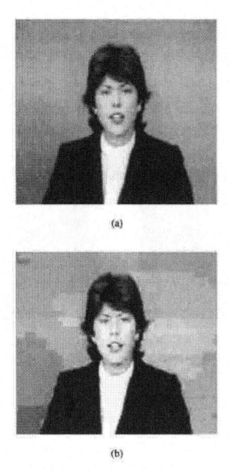

Figure 9. "Claire" at 0.19 bpp: (a) MVQ coder; (b) JPEG coder.

a property which is typical of a technique that operates in the subband domain. Blocked artifacts, particularly annoying at low bit rates, are visible from Figure 9b.

The effectiveness of the proposed technique for video coding is confirmed by the following experimental results. The image sequence is divided into Group of Pictures (GOP) that are 30 frames long: in particular, the first frame of each GOP is coded in intraframe mode. In addition, entropy coding of the motion vectors and of the segmentation mask is reinitialized at the beginning of each GOP. The total bit rate is given by the sum of the bits needed to code the frames and the bits needed for the motion information and the segmentation mask. Figure 10 shows the PSNR for the first GOP of the image sequence "Miss America," coded at a total bit rate of 128, 96 and 64 kbit/s: the original sequence is in CIF format at 30 frame/s and all the frames are coded. These results were obtained by setting the threshold T to the values 75, 110 and 230, respectively. In Fig. 11, we show the bit rate required

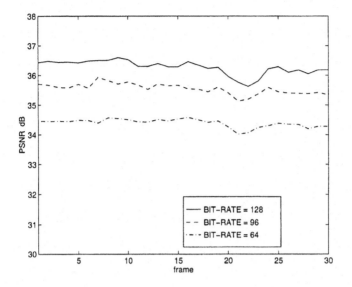

Figure 10. PSNR for the image sequence "Miss America."

to code each frame: this number does not include the information relative to the motion vectors and the segmentation masks, which together represent an average bit rate of 34 kbit/s (30, 24 kbit/s) out of the total bit rate of 128 kbit/s (96, 64 kbit/s).

Figure 14 shows the reconstructed frame number 8 of sequence "Miss America" at 128 kbit/s. Besides the good visual quality of each decoded frame, the sequence appears to have a good quality also when played on the video display.

Figure 12 shows frame 2 of the coded image with the motion vectors to predict from frame number 1. From the figure, it is clearly noticeable that the static background is correctly classified as not moving. This is particularly important for an effective coding of the motion information using a dictionary based coder: in fact, long runs of zeros and also similar patterns of motion vectors in successive frames can be very effectively coded by such a technique [30].

Figure 13a shows the segmentation mask identifying the 16×16 blocks in the original image where motion compensation from the previous frame is effective and below the threshold T_{MSE}. In Figure 13b we show how the mask is mapped into the multiresolution decomposition of the difference frame. The black region corresponds to blocks that are temporally predicted, while the white regions corresponds to blocks that are coded using the MVQ coder described in Section 3.

To further reduce the bit rate, it is possible to skip frames in the original image and interpolate them at the receiver. In the following, we present some preliminary results about this possibility. In particular, we applied the same coding algorithm described above to the even frames of the image sequence. At the decoder, the missing frames were built on the basis of motion information. Starting from the decoded frame n and the motion

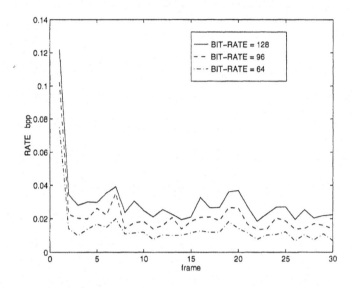

Figure 11. Bit rate for the first 30 frames of "Miss America."

Figure 12. Motion vectors for frame 2 of "Miss America."

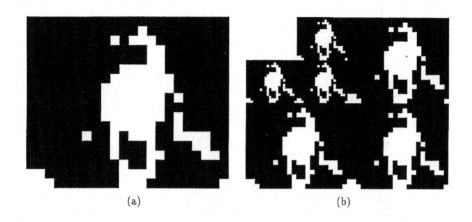

(a)   (b)

Figure 13. (a) Segmentation mask for "Miss America" and (b) its mapping to the multiresolution decomposition.

Figure 14. Decoded frame 8 of "Miss America" at 128 kbit/s.

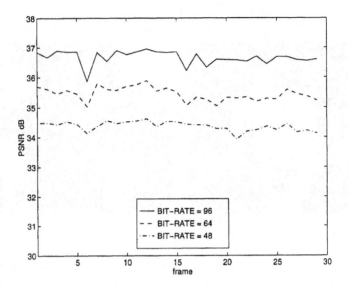

Figure 15. PSNR for the image sequence "Miss America" with interpolation at the receiver.

vector field $V(n)$ to predict frame $n + 2$, frame $n + 1$ is reconstructed at the decoder using a motion field $\hat{V}(n)$ that is obtained by dividing by 2 the components of each motion vector in $V(n)$. Moreover, the pixel intensities of each 16×16 block in the interpolated frame $n + 1$ are computed by averaging the values of the corresponding block in frame $n + 2$ and of the block in frame n addressed by the motion vector. Figure 15 shows the PSNR obtained using this simple strategy. The considered average bit rates are 96, 64 and 48 kbit/s, including 20, 18 and 16 kbit/s for motion information and the segmentation masks, respectively. The values of the threshold T used to obtain such coding rates were 60, 125 and 220. Figure 16 shows the interpolated frame number 7 and the decoded frame number 8 of the image sequence at 48 kbit/s.

6. Conclusions

In this paper, we presented a scheme for video coding that exploits temporal redundancy as well as intrinsic redundancy of the multiresolution decomposition of images. In particular, motion compensation is used to predict blocks of the current frame from blocks in the previously decoded frame. We classify the blocks in the current frame into two classes, based on the mean squared prediction error. The current frame is therefore segmented into two regions, corresponding to effective and non-effective motion compensation, respectively. Typically, one region corresponds to static background and to objects with approximately uniform motion: the other region, where motion compensation did not give satisfactory results, corresponds to objects with more complicated motion and to uncovered background.

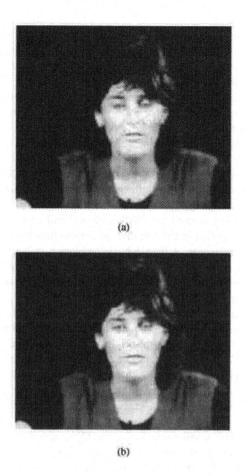

(a)

(b)

Figure 16. (a) Interpolated frame 7 and (b) decoded frame 8 of "Miss America" at 48 kbit/s.

The difference between the current frame and its motion compensated prediction is subsequently decomposed into a multiresolution pyramid of subbands. The coding algorithm exploits temporal redundancy by setting to zero the coefficients of the first two high resolution levels of the multiresolution decomposition that correspond to the region where motion compensation was effective. In this way, no coding is required for the high resolution residual details corresponding to the regions of the current frame where motion based prediction was effective. Nevertheless, the lower resolution levels of the subband pyramid are still coded, independently of the motion based segmentation mask. In this way, it is possible to correct for inaccurate motion estimation, while discarding details at high resolutions where the human visual system is less sensible.

The lower resolution subbands and the remaining coefficients in the higher resolution subbands of the difference images are coded using an efficient scheme that exploits interband redundancy of the multiresolution pyramid. In particular, high energy regions of subband coefficients, which typically correspond to edges in the original image, tend to have similar shapes and to be located in corresponding regions in all the subbands. Our coding scheme predicts blocks in one subband from blocks of the same dimension and located around the corresponding position in the lower resolution subband. Initial conditions for the prediction scheme are provided by coding with a lattice vector quantizer the lowest resolution subbands. A range block in one subband is therefore predicted from a domain block whose pixel positions can be shuffled and whose pixel intensities can be multiplied by a scaling factor before matching. Thus, a code for an entire block of coefficients is simply given by the address of the corresponding domain block, plus a code for a rotation and a scaling factor. Similarly to what is done in fractal coders, we allow for block splitting when the prediction from the lower resolution subband is not satisfactory. When the block dimension reaches the smallest size, the block is coded using PVQ.

Our coding scheme aims at exploiting both temporal and intra-frame redundancy and appears to be conceptually simple and effective, as demonstrated by experiments with still images and with video sequences. Although based on block matching and block prediction, the use of multiresolution coding inside the temporal prediction loop avoids blocked artifacts in the reconstructed sequences. However, ringing effects, typical of subband based coders, may be visible near edges, and the very detailed regions in the original image appear to be smoothed, due to inaccurate coding of the high frequency subbands. For video coding, reconstruction errors located in different positions of successive frames may appear as intensity changes that are not present in the original image. This kind of artifact is however less visible than in DCT based coders, where intensity changes are associated to block shaped regions. Coded sequences appear to have good visual quality, despite of the low bit rate. In addition, multiresolution coding allows for scalability of the bit stream.

The motion compensation procedure adopted for the proposed coder is very simple compared to more sophisticated solutions considered in the literature. The effectiveness of the complete video coding scheme relies upon the MVQ algorithm applied to difference images. In MVQ, the coding time is mainly devoted to the subband block search. Although the computation time is higher than that required by block based DCT coders, the reduced search area that we consider keeps the coding complexity at an acceptable level. At high rates, however, it can happen very often that interband prediction does not give good results. As a consequence, blocks are splitted, and the repetition of the search procedure for smaller blocks increases the coding time. Moreover, as the quality increases, the technique reduces to standard adaptive PVQ of the subbands. Nevertheless, at low bit rates, we find that the use of combined PVQ and interband prediction is beneficial.

We have also shown in the paper that it is possible to further reduce the bit rate for video coding by skipping frames at the transmitter and interpolating them at the receiver. We are currently investigating more sophisticated schemes for motion based segmentation and compensation: this would allow for more efficient coding of the difference frames and for content driven interpolation schemes at the receiver.

Acknowledgment

The authors would like to thank the people of the Image Processing Laboratory of the Dipartimento di Elettronica e Informatica for fruitful discussions and programming help.

References

1. H. Li, A. Lundmark, and R. Forchheimer, "Image Sequence Coding at Very Low Bitrates: A Review," *IEEE Trans. on Image Processing*, 1994, pp. 598–609.

2. H. G. Musmann, M. Hötter, and J. Ostermann, "Object-oriented analysis-synthesis coding of moving images," *Image Communication*, vol. 1, no. 2, 1989, pp. 117–138.

3. R. Rinaldo and G. Calvagno, "Image Coding by Block Prediction of Multiresolution Subimages," *IEEE Trans. on Image Processing*, vol. 4, 1995, pp. 909–920.

4. P. P. Vaidyanathan, *Multirate Systems and Filter Banks*, Prentice Hall, 1993.

5. J. W. Woods and S. D. O'Neil, "Subband coding of images," *IEEE Trans. on Acoustics Speech and Signal Processing*, 1986, pp. 1278–1288.

6. M. Vetterli and C. Herley, "Wavelets and Filter Banks—Theory and Design," *IEEE Trans. on Signal Processing*, 1992, pp. 2207–2232.

7. A. Cohen, I. Daubechies, and J. C. Feauveau, "Biorthogonal Bases of Compactly Support Wavelets," *Communications on Pure and Applied Mathematics*, 1992, pp. 485–560.

8. I. Daubechies, "The wavelet transform, time-frequency localization and signal analysis," *IEEE Trans. on Information Theory*, 1990, pp. 961–1005.

9. S. Mallat, "Multifrequency channel decomposition of images and wavelet models," *IEEE Trans. on Acoustics, Speech and Signal Proc.*, 1989, pp. 2091–2110.

10. E. P. Simoncelli and E. H. Adelson, "Subband Transforms," in J. W. Woods (ed.), *Subband Image Coding*, Kluwer Academic Publishers, 1991, pp. 143–192.

11. D. Esteban and G. Galand, "Application of Quadrature Mirror Filters to Split Band Voice Coding Schemes," in *Proc. IEEE ICASSP*, 1977, pp. 191–195.

12. J. D. Johnston, "A Filter Family Designed for Use in Quadrature Mirror Filter Banks," in *Proc. IEEE ICASSP*, 1980, pp. 291–294.

13. A. Pentland and B. Horowitz, "A Practical Approach to Fractal-Based Image Compression," in *Proc. Data Compression Conference*, 1991, pp. 176–185.

14. R. Rinaldo and A. Zakhor, "Inverse and Approximation Problem for Two-Dimensional Fractal Sets," *IEEE Trans. on Image Processing*, vol. 3, 1994, pp. 802–820.

15. J. M. Shapiro, "Embedded Image Coding Using Zerotrees of Wavelet Coefficients," *IEEE Trans. on Signal Processing*, vol. 41, 1993, pp. 3445–3462.

16. A. S. Lewis and G. Knowles, "Image Compression Using the 2-D Wavelet Transform," *IEEE Trans. on Image Processing 1(2)*, 1992, pp. 244–250.

17. M. Barnsley, *Fractals everywhere*, Boston: Academic Press, 1988.

18. T. R. Fischer, "A Pyramid Vector Quantizer," *IEEE Trans. on Information Theory*, 1986, pp. 568–583.

19. A. Gersho and R. M. Gray, *Vector Quantization and Signal Compression*, Boston, MA: Kluwer, 1992.

20. F. Bellifemine, C. Cafforio, A. Chimienti, and R. Picco, "Combining DCT and subband coding into an intraframe coder," *Signal Processing: Image Communication*, vol. 5, no. 3, 1993, pp. 235–248.

21. D. J. Sakrison, "A geometric Treatment of the Source Coding and Vector Quantization," *IEEE Trans. on Information Theory*, 1968, pp. 481–486.

22. T. M. Cover and J. A. Thomas, *Information Theory*, New York: Wiley, 1991.

23. M. E. Blain and T. R. Fischer, "A Comparison of Vector Quantization Techniques In Transform an Subband Coding of Imagery," *Signal Processing: Image Communication*, vol. 3, 1991, pp. 91–105.

24. N. S. Jayant and P. Noll, *Digital Coding of Waveforms*, Englewood Cliffs, NJ: Prentice Hall, 1984.

25. A. E. Jacquin, "Image Coding based on a Fractal theory of Iterated Contractive Image Transformations," *IEEE Trans. on Image Processing*, 1992, pp. 18–31.

26. Y. Fisher, E. W. Jacobs, and R. D. Boss, "Fractal Image Compression Using Iterated Transforms," in J. A. Storer (ed), *Image and text compression*, Dordrecht, Netherlands: Kluwer Academic Publishers, 1992, pp. 35–61.

27. R. H. Bamberger, S. L. Eddins, and V. Nuri, "Generalized Symmetric Extension for Size-Limited Multirate Filter Banks," *IEEE Trans. on Image Processing*, 1994, pp. 82–87.

28. M. Nelson, *The Data Compression Book,* M&T Books, 1992.

29. G. K. Wallace, "The JPEG Still Picture Compression Standard," *Communications of the ACM*, 1991, pp. 30–44.

30. J. Ziv and A. Lempel, "A universal algorithm for sequential data compression," *IEEE Trans. on Information Theory,* vol. IT-23, 1977, pp. 337–342.

Multidimensional Systems and Signal Processing, 8, 151–184 (1997)

Multiscale, Statistical Anomaly Detection Analysis and Algorithms for Linearized Inverse Scattering Problems

ERIC L. MILLER   elmiller@cdsp.neu.edu
The Communications and Digital Signal Processing Center, Department of Electrical and Computer Engineering,
235 Forsyth, Northeastern University, 360 Huntington Ave., Boston, MA 02115

ALAN S. WILLSKY
Laboratory for Information and Decision Systems, Department of Electrical Engineering and Computer Science,
Massachusetts Institute of Technology, Cambridge, Massachusetts 02139

Abstract. In this paper we explore the utility of multiscale and statistical techniques for detecting and character-izing the structure of localized anomalies in a medium based upon observations of scattered energy obtained at the boundaries of the region of interest. Wavelet transform techniques are used to provide an efficient and physically meaningful method for modeling the non-anomalous structure of the medium under investigation. We employ decision-theoretic methods both to analyze a variety of difficulties associated with the anomaly detection problem and as the basis for an algorithm to perform anomaly detection and estimation. These methods allow for a quanti-tative evaluation of the manner in which the performance of the algorithms is impacted by the amplitudes, spatial sizes, and positions of anomalous areas in the overall region of interest. Given the insight provided by this work, we formulate and analyze an algorithm for determining the number, location, and magnitudes associated with a set of anomaly structures. This approach is based upon the use of a Generalized, M-ary Likelihood Ratio Test to successively subdivide the region as a means of localizing anomalous areas in both space and scale. Examples of our multiscale inversion algorithm are presented using the Born approximation of an electrical conductivity problem formulated so as to illustrate many of the features associated with similar detection problems arising in fields such as geophysical prospecting, ultrasonic imaging, and medical imaging.

Key Words:

1. Introduction

The goal of many applied problems is the recovery of information regarding the structure of a physical medium based upon measurements of scattered radiation collected at the bound-aries [8, 15, 19, 43, 48]. For some of these tomographic-type inverse problems, one seeks a complete description (in the form of an image in two dimensions or a volumetric rendering in 3D) of the structure of the medium. In other cases, however, the full reconstruction is not needed; rather, the ultimate objective is to extract the structure of areas in the medium which are, in some sense, anomalous; that is, regions where the nature of the medium differs from some prior set of expectations. This *anomaly detection problem* arises, for example, in geophysical prospecting where in many instances the fundamental issue is the determi-nation of oil bearing regions in the earth and medical imaging where tumor detection is of import.

As discussed in [27, 29, 31, 33, 34, 44, 45] for many of the application areas previously cited, methods for solving the anomaly detection problem typically proceed by initially generating the full, pixel-by-pixel reconstruction and subsequently post-processing the results to determine the nature of anomalous structures. The necessity of generating a solution to the so-called "full inverse problem" however makes these schemes rather unattractive. Indeed, for many interesting applications, obtaining a full reconstruction of the medium presents a collection of well-known and extensively studied challenges [2, 3, 40] which suggest that solving this problem as the first step toward localizing anomalies should be avoided. In this paper we demonstrate the utility of a multiscale framework for explicitly solving the spatial anomaly detection problem in the context of linearized inverse scattering (also known as diffraction tomography [15]) applications.

The basis for solving the anomaly detection problem is the use of wavelet transforms and the statistical theories of optimal estimation and detection to develop both efficient algorithms for anomaly detection and localization and analytical insight into the nature of the problem and the limits of performance that result from the fundamental physics relating the characteristics of the medium to the observations. In [39, 40], we introduced the use of wavelet transforms and multiresolutional statistical techniques for overcoming many of the challenges associated with the solution of full reconstruction, linearized inverse electrical conductivity problems. Many of the results in [39, 40] followed from the use of multiscale, statistical regularization methods for the incorporation of prior knowledge into the inversion routine. The use of such prior statistical models automatically implies an assumption of some type of statistical regularity on the field and therefore fails to capture adequately the presence of anomalies or localized inhomogeneities. Thus, roughly stated, the problem considered in this paper is the detection, localization, and estimation of such anomalies superimposed on a background of know statistical structure and observed indirectly through the scattering measurements.

The consideration of the anomaly detection problem raises a variety of questions beyond those arising in the full reconstruction inverse problem. How many anomalies are there? Where are they located? What are their sizes? What are their amplitudes? Given answers to the first three of these problems, the fourth is a variant of the full inverse problem in which we focus our attention on determining the magnitudes of only the previously identified anomalous regions. The determination of the number, sizes and locations of the anomalous regions is, however, a potentially daunting collection of tasks as a result of the vast number of combinations of anomaly structures which, in principle, must be explored in the generation of a solution.

Over the past decade, significant work has been performed in the area of anomaly detection from tomographic-type measurements. In [44], Rossi and Willsky were concerned primarily with the use of estimation-theoretic analysis and algorithmic methods for determining the location of a single object of known size and structure given noisy and sparse computed tomography (CT) measurements. Recently, these results have been extended by Devaney and co-workers [16, 17, 46] in consideration of diffraction tomography (DT) and exact scattering applications. More closely related to the problem of interest in this paper is the work of Bresler, Fessler and Macovski. In [5], the authors examined a 3D reconstruction problem from CT measurements in which the first step of their algorithm required the localization of an unknown number of anomalies of unknown structure. The solution to this problem presented in [5] was to estimate the required parameters for a pre-determined, maximum number of anomalies knowing that further processing would eliminate falsely identified anomalous regions.

In this paper, we present a scale-recursive algorithm for anomaly detection and characterization given DT-type data. Here, the tools of optimal hypothesis testing are used to make a sequence of anomaly detection and localization decisions starting at coarse scales, thereby allowing for the detection of spatially large anomaly structures and providing coarse localization of finer scale anomalies, and then moving to finer ones. This algorithm is significant for two reasons. First, this approach provides a computationally efficient and accurate means of localizing areas of anomalous behavior. Second, the anomaly characterization algorithm may be viewed as a highly efficient first stage in a larger image processing application. Specifically, the output of the algorithm could be refined (for example via the methods described in [5, 44] generalized to the case of diffraction tomography) by higher level processing stages concerned with issues such as identification, classification, or imaging. Toward this end, in Section 6.3, we present one way in which knowledge of the anomaly structures can be used to supplement the information in the prior statistical model in order to improve the output of a least-squares, pixel-by-pixel reconstruction of the region of interest.

In addition to the development of the scale-recursive processing algorithm, by using these same statistical techniques, we provide analysis of the anomaly detection problem that not only yields overall performance limits, but also guides the detection procedure. For example, we are able to define and determine the statistical distinguishability of a small scale, large amplitude anomaly from a larger scale, but smaller magnitude structure or a pair of closely spaced anomalies from a single, broader anomalous region. The use of the results from this analysis can then tell us at what scale and in which regions to terminate our detection procedure, i.e. when finer scale localization is unwarranted given the available data.

In Section 2, we present an overview of the particular anomaly detection problem of interest in this work. The formal definition of the anomaly detection problem as one of optimal hypothesis testing and a review of results from statistical decision theory is provided in Section 3. In Section 4 we demonstrate the utility of our framework in characterizing the *detectability* of an anomaly. Section 5 is devoted to the question of the *distinguishability* of anomalies as a function of their relative positions and structures. In Section 6 we develop and analyze a scale-recursive algorithm for anomaly detection, localization, and estimation, and present the results of its performance under a variety of experiment conditions. Conclusions reached in this paper and directions for further work are presented in Section 7.

2.   A Multiscale Framework for Inverse Scattering

2.1.   The Scattering Problem

The context in which we develop our anomaly detection algorithm is a low-frequency, two-dimensional inverse electrical conductivity problem illustrated in Figure 1 and similar to problems arising in the field of geophysical prospecting [23, 24, 48] and medical imaging using electrical impedance tomography [18–20, 22, 29–31, 43]. Here, we have an array of

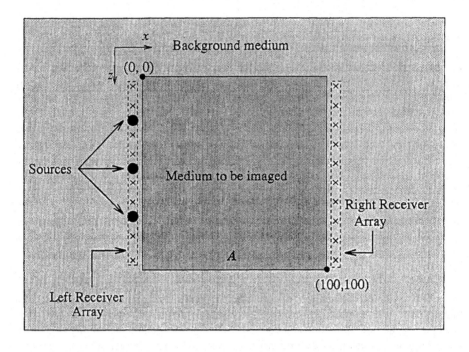

Figure 1. Configuration of inverse conductivity problem. The electromagnetic sources (indicated by the black circles) emit time-harmonic waves into a lossy medium which subsequently are scattered by conductivity inhomogeneities located in the darkly shaded rectangle, A. The secondary fields are observed at one or both receiver arrays located on either vertical edge of the region under investigation. Based upon these observations, the objective of the inverse problem is the reconstruction of the conductivity perturbation.

electromagnetic line-sources oriented perpendicularly to the page emitting time-harmonic, waves into a lossy medium. The electrical properties of this environment are assumed to be decomposed into the sum of an infinite, known, and constant background and a conductivity perturbation, g, with support restricted to region A in Figure 1. The fields from the transmitters are scattered by g, and the secondary fields are observed at one or both of the receiver arrays positioned on the vertical edges of region A. Based upon these observations, the objective of the problem is to detect and localize areas in the region of interest where the structure of g is, in a sense to be defined below, anomalous.

We consider the collection of eighteen scattering experiments defined in Table 1 where each such experiment produces a vector of measurements comprised of the in-phase and quadrature components of the observed scattered field obtained over one of the two receiver arrays due to energy put into the medium from one of the sources operating at a particular frequency. As is shown in [9], the use of the first Born approximation yields the following linear relationship between the vector of observations associated with the ith scattering

Table 1. Data set definitions for observation processes of interest in the paper. The notation $x : y : z$ indicated that the sources are distributed in y increments along a line from x to z.

Experiment number	Source Position	Frequency of source (Hz)	Receiver Array
1 – 6	0:20:100	$f_{HI} = 10,000$	Left
7 – 12	0:20:100	$f_{MID} = 1,000$	Left
13 – 18	0:20:100	$f_{LO} = 100$	Right

experiment, y_i, and a discrete representation of the two dimensional conductivity anomaly, g

$$y_i = T_i g + n_i \quad i = 1, 2, \ldots, 18 \tag{1}$$

where the matrices T_i encompass the (linearized) physics and n_i is an additive, zero-mean, uncorrelated, random vector representing the noise in the data. That is, the ith noise is modeled as $n_i \sim \mathcal{N}(0, r_i I)$ where I is an appropriately sized identity matrix.[1] The discrete representation of the conductivity g is constructed using the so-called "pulse" set of basis functions where the conductivity is assumed to be piecewise constant over an $N_{g,x} \times N_{g,z}$ grid of square pixels covering A [26]. For future reference, we define the "stacked" system of data

$$y = Tg + n \tag{2}$$

where $y^T = [y_1^T \, y_2^T \, \ldots \, y_{18}^T]$ with T and n defined accordingly.

2.2. A Multiscale Representation of the Problem

The detection techniques developed in Sections 4–6 are based upon a linear model relating multiresolution representations of g and n_i to a multiresolution representation the data, y_i. A scale-space representation of the problem has been chosen for two reasons. First, the matrices T_i in (1) are of the class which are made sparse in the wavelet transform domain [1, 4] thereby lowering the computational complexity of the detection algorithm in Section 6. Although not considered extensively in this work, such computational benefits are explored in [41]. Second, as we discuss below, a collection of useful and physically meaningful models for the non-anomalous behavior of the conductivity field are specified easily in the wavelet domain.

Following the work in [39, 40], orthonormal, discrete wavelet transform (DWT) [14] operators (matrices) $\mathcal{W}_i$ and $\mathcal{W}_g$ are used to move from physical to scale space in the following manner

$$\eta_i = \mathcal{W}_i y_i = (\mathcal{W}_i T_i \mathcal{W}_g^T)(\mathcal{W}_g g) + \mathcal{W}_i n_i \equiv \Theta_i \gamma + \nu_i \tag{3}$$

where $\mathcal{W}_g^T \mathcal{W}_g = \mathcal{W}_i^T \mathcal{W}_i = I$ follows from the orthonormality of the wavelet transformation [14, 35]. There are a variety of reasons why we may wish to use different transforms
for the data than for g. First, from Figure 1, each data set is to be collected over a 1D array
of receivers. Hence, $\mathcal{W}_i$ will act on a one dimensional signal while $\mathcal{W}_g$ is used to transform
the 2D conductivity profile. Additionally, it may be the case that the lengths of each data
record vary from one observation process to the next. Finally, analogously to the physical
space case, we define the stacked systems

$$\eta = \Theta\gamma + \nu \tag{4}$$

where $\eta = [\eta_1^T \; \eta_2^T \; \ldots \eta_{18}^T]^T$, Θ and ν are defined analogously and $\nu \sim \mathcal{N}(0, R)$ with
$R = diag(r_1 I, r_2 I, \ldots, r_{18} I)$.

2.3. Multiscale Prior Models

Recently there has been significant work in the use of fractal models for describing the
spatial distribution of geophysical quantities. In [13], Crossley and Jensen explore the
propagation of acoustic radiation in the Earth's crust using a velocity model composed of
the sum of a deterministic profile and a fractal perturbation. In considering the distribution
of hydraulic conductivity, Brewer and Wheatcraft [6] employ a wavelet-based model very
similar to the one described below as a means of interpolating coarse scale observations of
hydraulic conductivity to finer scales. Brown [7] relates both the electrical and hydraulic
conductivities in the earth to a self-similar model for the height distribution in rock fractures
and studies the resulting fluid and current flow patterns though such a formation. Finally,
the propagation of electromagnetic radiation through media with fractal characteristics has
been studied extensively by Jaggard and co-workers [32].

With this work as motivation, we use a stochastic, fractal-type model to describe the
spatial distribution of the electrical conductivity in the absence of anomalies. While there
are many self-similar models which may be used to describe the conductivity, results of
Wornell [50], Tewfik [47], and Chou et al. [10–12] suggest that there exist a wide range of
statistical models specified *directly* in the wavelet transform domain possessing the desired
modeling characteristics and simple structures thereby making them quite attractive for use
in signal and image processing applications.

Under the particular wavelet-based model of interest in this paper, the wavelet coefficients of the non-anomalous conductivity field, denoted by the vector $\tilde{\gamma}$, are taken to be
uncorrelated, Gaussian random variables. That is, $\tilde{\gamma}$ is distributed according to

$$\tilde{\gamma} \sim \mathcal{N}(0, P_0) \tag{5}$$

where P_0 is a *diagonal* matrix whose nonzero entries are the variances of the corresponding
wavelet coefficients. While a detailed description of the internal structure of P_0 is presented
in [35, 50], the fractal-type behavior of the process is obtained by taking the variance of
the wavelet coefficients to vary exponentially with scale. Coefficients in $\tilde{\gamma}$ governing the
coarsest scale behavior of the conductivity have relatively large variances while fine scale
components possess smaller variances.

3. Anomaly Detection as a Hypothesis Testing Problem

3.1. A Model for the Conductivity

The objective of the anomaly detection problem is to determine those areas in A where the behavior of g is anomalous in that in these regions g differs from some prior set of beliefs regarding the manner in which the conductivity is expected to behave. Thus, the conductivity g is decomposed as

$$g = \tilde{g} + \bar{g} \tag{6}$$

where $\tilde{g}$ represent that portion of g consistent with our prior assumptions and $\bar{g}$ encompasses the anomalous behavior of the conductivity; that is, the perturbation of the conductivity away from its non-anomalous structure. In the wavelet transform domain, (6) takes the form

$$\gamma = W_g g = W_g \tilde{g} + W_g \bar{g} \equiv \tilde{\gamma} + \bar{\gamma}. \tag{7}$$

As will be seen in Sections 4–6, considerable insight into the anomaly detection problem is obtained through performance analysis carried out using anomaly structures of varying sizes (i.e. spatial scales) located in different regions of A. Also, the primary intent of the detection algorithm presented in Section 6 is to localize quickly and efficiently regions where anomalies are suspected to exist. As region A is pixelated into an $N_{g,x} \times N_{g,z}$ grid and because we perform anomaly localization through a process of spatial subdivision, we are lead naturally to consider a representation in which anomalous regions are defined to be superpositions of rectangular subsets of A.

Referring to Figure 2, the structure of the ith anomaly in A is defined by its magnitude, a_i, its size, and its location in A. The area of an anomaly defines its *scale* in that small scale anomalies are correspondingly small in area and similarly for larger scale anomaly structures. Mathematically, the form for the anomalous behavior of the conductivity over the region A is

$$\bar{g} = \sum_{j=1}^{N_a} b_j a_j = Ba. \tag{8}$$

Here, N_a is the number of anomalous regions located in A, a_j is a scalar defining the magnitude of each anomaly, and b_j represents the discrete indicator function over the jth rectangular region in $\bar{g}$. In (8), the column vector a represents the collection of anomaly amplitude coefficients while B is the matrix whose j^{th} column is b_j. In the wavelet transform domain, (8) is written as

$$\bar{\gamma} = \sum_{j=1}^{N_a} (W_g b_j) a_j \equiv \mathcal{B}a \tag{9}$$

where $\mathcal{B} = [W_g b_1 \ W_g b_2 \ \ldots \ W_g b_{N_a}]$. Finally, use of (7) and (9) in (4) yields the following

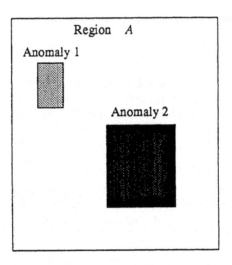

Figure 2. General structure of anomalous regions of interest in this paper. The magnitudes, a_1 and a_2 of the two anomalies shown here are proportional to the color of the corresponding rectangles.

relationship among the anomaly structures, the non-anomalous background $\tilde{g}$ or $\tilde{\gamma}$, and the data

$$\eta = \Theta\bar{\gamma} + \Theta\tilde{\gamma} + \nu = \Theta\mathcal{B}a + \Theta\tilde{\gamma} + \nu \tag{10}$$

where, because $\tilde{\gamma}$ and ν are taken to be uncorrelated,

$$P_\eta = E[\eta\eta^T] - E[\eta]E[\eta^T] = \Theta P_0 \Theta^T + R. \tag{11}$$

Note that the analysis methods and algorithmic techniques presented in this work are based entirely on an observation model of the form in (10). In particular, the results in this paper are not dependent upon the assumption of rectangular anomalies; rather structures with arbitrary shapes and orientations can be employed in principle through the appropriate specification of the matrix $\mathcal{B}$. Nonetheless, as will be seen in Sections 4–6 of this paper, rectangular structures prove to be highly useful for obtaining significant insight into the nature of the anomaly characterization problem and as the basis for an algorithm designed to extract this information from observed scattered fields.

To provide a normalized notion of the overall size of an anomaly, we define an SNR-type quantity called the *anomaly-to-background ratio* (ABR) which provides a measure of the energy in an anomaly relative to that of $\tilde{g}$. Mathematically, we have for an anomaly $\bar{g}$ composed of a single rectangular region defined by the column vector b and with amplitude a

$$ABR^2 = \frac{\text{Power in } \bar{g}}{\text{Expected power in } \tilde{g}} = \frac{a^2(b^T b)}{tr(\bar{P}_0)} \tag{12}$$

where $tr(M)$ is the trace of the matrix M and $\bar{P}_0 = \mathcal{W}_g^T P_0 \mathcal{W}_g$ is the covariance matrix of $\tilde{g}$.

As described in [40], under the Born approximation used to obtain (1), $g = \mathcal{W}_g^T \gamma$ represents a perturbation about a known, constant background conductivity, g_0. From physical principles, the overall conductivity, $g_0 + g = g_0 + \tilde{g} + Ba$ must be greater than zero. Thus, in theory the elements of a may assume both positive as well as negative values so long as the positivity constraint is satisfied. To simplify matters, in this paper we assume that the a_i are strictly greater than zero corresponding to regions of locally higher conductivity than the background.

3.2. The M-ary Hypothesis Testing Problem

In Section 6, we consider a statistical decision-theoretic methodology for reconstructing $\bar{\gamma}$ which is based upon a sequence of M-ary Generalized Likelihood Ratio Tests (GLRT) as a means of localizing an unknown number of anomalous regions in A. The mathematical description of each such test begins with the formulation of the following M hypotheses, H_i for $i = 0, 1, 2, \ldots, M - 1$, corresponding to M different configurations of anomalous areas

$$H_i : \quad \eta = \Theta \mathcal{B}_i a_i + \Theta \tilde{\gamma} + \nu \quad i = 0, 1, 2, \ldots, M - 1. \tag{13}$$

Note that from (13) under H_i we have, $\eta \sim \mathcal{N}(\Theta \mathcal{B}_i a_i, P_\eta)$ where P_η is given by (11).

The hypothesis test is implemented as a rule which when given the data, indicates which of the H_i is true. Because it will be the case in Section 6 that the a_i are taken to be deterministic but unknown parameters, a standard likelihood ratio test solution to the hypothesis testing problem [49] cannot be employed in this context. Rather, we use a Generalized Likelihood Ratio Test (GLRT) [49] for performing the test. This procedure requires first that an estimate of each a_i be computed assuming that H_i is correct. As this problem is, in general, ill-posed, we choose here to use the following regularized, least squares estimate

$$\hat{a}_i = (\mathcal{B}_i^T \Theta^T P_\eta^{-1} \Theta \mathcal{B}_i + \alpha I)^{-1} \mathcal{B}_i^T \Theta^T P_\eta^{-1} \eta. \tag{14}$$

where the parameter α is used to control the degree of regularization.

Given $\hat{a}_i$, the hypothesis testing rule employed in this paper is

$$\text{Choose } H_i \text{ with } i = \begin{cases} 0 & \max_j L_j(\eta) < 0 \\ \arg\max_j L_j(\eta) & \text{otherwise} \end{cases} \tag{15}$$

where

$$L_j(\eta) = l_j(\eta) - l_0(\eta) \quad j = 1, 2, \ldots, M - 1 \tag{16}$$

and for $j = 0, 1, 2, \ldots, M - 1$

$$l_j(\eta) = \eta^T P_\eta^{-1} \Theta \mathcal{B}_j \hat{a}_j - \frac{1}{2} \hat{a}_j^T \mathcal{B}_j^T \Theta^T P_\eta^{-1} \Theta \mathcal{B}_j \hat{a}_j. \tag{17}$$

3.3. The Binary Hypothesis Testing Case

While the algorithm for extracting anomaly information is based upon the M-ary GLRT, much of the analysis of the anomaly detection problem is performed in the context of the *binary hypothesis testing* (BHT) framework in which two alternatives, $\bar{\gamma}_0 = \mathcal{B}_0 a_0$ and $\bar{\gamma}_1 = \mathcal{B}_1 a_1$, are compared.[2] Traditionally, the analysis of the BHT centers around the probability of detection, P_d and the false alarm probability, P_f. For the linear-Gaussian model considered in this work, it is shown in [49] that P_d and P_f are related to the various quantities defining the structure of the problem via

$$d = \text{erfc}_*^{-1}(P_f) - \text{erfc}_*^{-1}(P_d) \tag{18}$$

where

$$d^2 = (\bar{\gamma}_1 - \bar{\gamma}_0)^T \Theta^T P_\eta^{-1} \Theta (\bar{\gamma}_1 - \bar{\gamma}_0) \tag{19}$$

$$\text{erfc}_*(x) = \int_x^\infty \frac{1}{\sqrt{2\pi}} e^{-t^2/2} dt. \tag{20}$$

Thus, based upon (18), we see that our ability to distinguish between two anomaly structures is intimately related to the Fisher discriminant, d, which has the interpretation of a "signal-to-noise" ratio [49]. Note that for a given P_f, larger d results in larger P_d and therefore better performance.

From (19) we observe that the performance of the binary hypothesis test is a function of both the geometric configurations, as captured in the matrices $\mathcal{B}_i$, and the magnitudes, a_i, of the two candidate anomaly structures. To better understand the role of these two factors, consider the case in which $\bar{\gamma}_i$ corresponds to a single rectangular region so that each $\mathcal{B}_i$ is a column vector and each a_i is a scalar. Substituting (9) into (19) and expanding the quadratic yields

$$\delta_1^2 a_1^2 - 2\delta_{1,0} a_1 a_0 + \delta_0^2 a_0^2 - (\text{erfc}_*^{-1}(P_f) - \text{erfc}_*^{-1}(P_d)) = 0 \tag{21}$$

where

$$\delta_j^2 = \mathcal{B}_j^T \Theta^T P_\eta^{-1} \Theta \mathcal{B}_j \quad \text{for } j = 0, 1 \tag{22}$$

$$\delta_{1,0} = \mathcal{B}_1^T \Theta^T P_\eta^{-1} \Theta \mathcal{B}_0. \tag{23}$$

In [37], it is shown that when viewed as a function of a_0 and a_1, (21) defines an ellipse the form of which is illustrated in Figure 3.[3] This ellipse indicates that, given the geometry of the candidate anomalies, $\mathcal{B}_0$ and $\mathcal{B}_1$, there are only certain combinations of a_0 and a_1 which will result in performance below that level dictated by a particular P_d and P_f. In fact, these points are precisely those that lie inside the plotted ellipse. Also, there exists a minimum level, $a_{1,0}^{min}$ (depending on the geometric structures of *both* anomalies) such that for $\bar{\gamma}_1 = \mathcal{B}_1 a_1$ with $a_1 > a_{1,0}^{min}$, the binary hypothesis test will achieve or exceed the P_d and P_f performance figures *independent* of a_0.

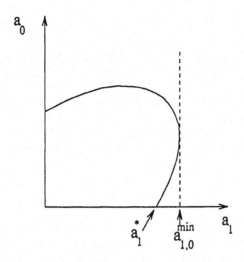

Figure 3. The structure of the ellipse defined by (21). The axes represent the magnitudes of anomaly structures in a binary hypothesis testing problem. As discussed in Section 2, a_0 and a_1 are taken to be nonnegative so that only the first quadrant is shown in this illustration. Here a_1^* is the minimum amplitude of $\bar{\gamma}_1$ required to detect this structure when the alternate hypothesis is $\bar{\gamma}_0 = 0$ for a BHT with prespecified P_d and P_f. The value $a_{1,0}^{min}$ is the minimum intensity of $\bar{\gamma}_1$ required to ensure that for *any* $\bar{\gamma}_0$ the performance of the resulting BHT meets or exceeds that defined by P_d and P_f.

4. Detectability Analysis

The first issue we address in conjunction with the anomaly detection problem is that of the detectability of an anomaly as a function of location, spatial size, and amplitude. After defining a particular collection of anomaly structures, a set of binary hypothesis testing problems are explored in which H_0 corresponds to there being no anomaly in the region while under H_1, a particular member of our anomaly collection is assumed to be present. The objective of the detectability analysis is to determine the minimum magnitude each such structure must possess to guarantee a prespecified level of performance from the binary hypothesis test.

Detectability is of interest due to the physics governing the relationship between the observations, η, and the conductivity, γ and the constrained experimental conditions in which data are collected only along the vertical edges of A. From these facts, it is not expected that arbitrarily small (in scale and magnitude) anomalies will be detectable with arbitrary precision throughout A. Rather, we anticipate that small anomalies should be readily detected only close to the observation points while interior to region A small scale structures would require significantly larger magnitudes to be as detectable as their counterparts closer to the edges.

With this intuition in mind, we consider a family of anomaly structures generated by a set of dyadic tesselations of A. For example, with $N_{g,x} = N_{g,z} \equiv N_g = 16$, we take as $\mathcal{J}_1$

the set of N_g^2 indicator functions which are one over single pixels in A and zero elsewhere. Analogously, $\mathcal{J}_2$ is the collection of $N_g^2/4$ characteristic functions over disjoint 2×2 sized regions of A. Thus, in general $\mathcal{J}_m$ (for m an integral power of 2) is the set of $(N_g/m)^2$ non-overlapping square regions of size $m \times m$ completely covering A. Finally, we define $\mathcal{J}$ as the union of all $\mathcal{J}_m$.

To begin our analysis of detectability, for each anomaly structure in $\mathcal{J}$, we consider a collection of binary hypothesis testing problems where the two hypotheses in the jth problem correspond to the situations in which no anomaly is present in A or a scaled version of the jth element of $\mathcal{J}$ is in A. Recalling (13), these alternatives take the form

$$H_0: \quad \eta = \Theta \tilde{\gamma} + \nu \tag{24a}$$

$$H_{1,j}: \quad \eta = \Theta \mathcal{B}_j a_j + \Theta \tilde{\gamma} + \nu. \tag{24b}$$

The goal of our detectability analysis then is to determine for each anomaly structure in $\mathcal{J}$, the minimum value of a_j, denoted a_j^*, such that the above hypothesis test attains a certain level of performance as specified by P_d and P_f.

The primary quantity used to characterize the performance of the binary hypothesis test in (24a)–(24b) is the Fisher discriminant discussed in the previous section which here takes the form

$$d_j^2 = a_j^2(\mathcal{B}_j^T \Theta^T P_\eta^{-1} \Theta \mathcal{B}_j) \equiv a_j^2 \delta_j^2 \tag{25}$$

where δ_j^2 is defined in (22) and represents the Fisher discriminant for the unit amplitude anomaly over the jth member of $\mathcal{J}$. Now, for a given P_d and P_f, (18) and (25) are combined to give the following expression for a_j^*:

$$a_j^* = \frac{\text{erfc}_*^{-1}(P_f) - \text{erfc}_*^{-1}(P_d)}{\delta_j}. \tag{26}$$

In Figure 4, a_j^* are plotted for all anomalies in $\mathcal{J}$ for the case in which data from the 18 experiments described in Table 1 at an SNR of 10 are available and where P_d is set to 0.95 and P_f is 0.05. In this work, the SNR associated with the anomaly-free observation process $\eta_i = \Theta_i \tilde{\gamma} + \nu_i$ with $\nu_i \sim (0, r_i^2 I)$ and $\gamma \sim (0, P_0)$ is defined as

$$SNR_i^2 = \frac{\text{Power per pixel in} \Theta_i \gamma}{\text{Power per pixel in } \nu_i} = \frac{tr(\Theta_i P_0 \Theta_i^T)}{N_g r_i^2}. \tag{27}$$

Thus, each 1×1 pixel in Figure 4(a) corresponds to an anomaly in $\mathcal{J}_1$ with the intensity of that pixel proportional to a_j^*. In all four cases, we see that near the middle of the region, the magnitude required to obtain the desired level of performance in the binary hypothesis test is significantly larger than that required near the vertical edges i.e. where the sources and receivers are located. For vertical values roughly in the range $40 \leq z \leq 60$, this effect is somewhat smaller. Also, as the areas of the anomalies increase, the required magnitudes decrease. This coincides with the intuition that large scale structures should be easier to

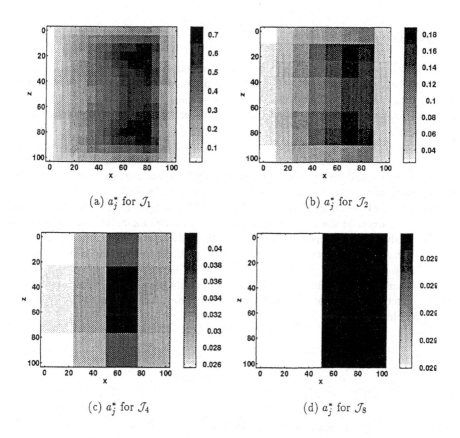

Figure 4. Value of a_j^* for all anomaly structures in $\mathcal{J}$ where the data from the experiments described in Table 1 at an SNR of 10 are used as input to the likelihood ratio test. Here, we have $P_d = 0.95$ and $P_f = 0.05$. Note that the scales in these images are all different with a^* decreasing significantly as the size of the anomalies increases.

detect than their fine scale counterparts. Finally, the ABR values in Figure 4 are quite small with the median values all less than 0.9. This implies that our statistical approach toward anomaly detection should prove quite advantageous in detecting relatively small amplitude conductivity anomalies.

To explain the behavior of a_j^*, we note that as described in [40] the low and medium frequency kernels are most sensitive to the conductivity structure over the horizontal range $0 \leq x \leq 50$ so that the required magnitude for an anomaly to be "seen" in this area should be relatively low. The smaller values of a_j^* in the region $40 \leq z \leq 60$ are due primarily to the combined coverage of this region by more observation kernels, T_i, than is the case for the top and bottom edges.

5. Distinguishability Analysis

In this section, we explore issues associated with our ability to successfully distinguish between pairs of candidate anomalies in order to obtain quantitative insight into the ambiguity
which exists in attempting to differentiate between anomalous structures of differing sizes,
locations, and magnitudes. The results of this work then are used both in the formulation
as well as the analysis of the detection algorithm in Section 6.

Before proceeding with the analysis, we note that the issue of distinguishability has been
considered previously in the context of electrical impedance tomography [18, 22, 29]. In
that work, distinguishability was examined in a deterministic setting where observation
perturbation was modeled as a bounded but otherwise unknown signal. Under such a
model, two conductivity profiles were defined to be distinguishable if the norm of the
difference between the data sets produced by each exceeded the noise level. The notion
of distinguishability developed below is rather different as it rests upon a statistical model
for both the additive measurement noise and background perturbations in the medium's
conductivity.

The mathematical formulation of the distinguishability problem of interest in this work
follows directly from Section 3.3. We begin by considering the following binary hypothesis
testing problem

$$H_j : \quad \eta = \Theta \mathcal{B}_j a_j + \Theta \tilde{\gamma} + \nu \tag{28a}$$

$$H_i : \quad \eta = \Theta \mathcal{B}_i a_i + \Theta \tilde{\gamma} + \nu. \tag{28b}$$

The primary tool for our distinguishability analysis is the quantity $a_{i,j}^{min}$ defined in Section 3.3
to be the smallest value of a_i such that the performance of the binary hypothesis test in (28a)–
(28b) meets or exceeds that defined by $P_{d,i,j}$ and $P_{f,i,j}$ independent of the amplitude of a_j.
Finally, for all experiments and for all i and j of interest in this section, $P_{d,i,j}$ is equal to
0.95 and $P_{f,i,j} = 0.05$.

In Figure 6, $a_{i,j}^{min}$ is shown as a function of $j \in \mathcal{J}$ in the case where the geometric structure
of anomaly $\tilde{\gamma}_i$ is given in Figure 5(a). Similarly, $a_{i,j}^{min}$ is displayed for the anomaly geometry
of Figure 5(b) in Figure 7. Essentially these two examples demonstrate the manner in which
the ability to differentiate structures is dependent upon the spatial position of the anomalies
in region A. In both cases, we see that the largest values of $a_{i,j}^{min}$ are associated with
hypothesis tests in which $\tilde{\gamma}_i$ is compared to a second, relatively close-by anomaly structure;
however, these amplitudes are roughly twice as large for the structure located toward the
middle of the region than for the anomaly closer to the source/receiver arrays.

In Table 2, the ABRs corresponding to the largest and smallest values for $a_{i,j}^{min}$ in Figures 6
and 7 are shown. That is for i fixed, the entries in the first column of Table 2 are the anomaly-
to-background ratios generated by

$$a_i^{max,min} = \max_j a_{i,j}^{min}$$

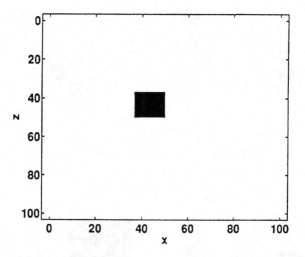

(a) First anomaly structure to be analyzed in distinguishability problems

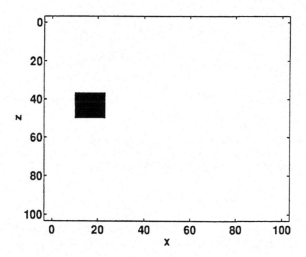

(b) Second anomaly structure to be analyzed in distinguishability problems

Figure 5. Anomaly structures to be analyzed in distinguishability problems

while those of the second column are associated with

$$a_i^{min,min} = \min_j a_{i,j}^{min}.$$

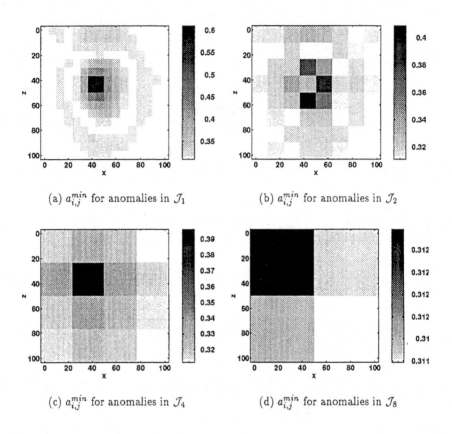

(a) $a_{i,j}^{min}$ for anomalies in $\mathcal{J}_1$ (b) $a_{i,j}^{min}$ for anomalies in $\mathcal{J}_2$

(c) $a_{i,j}^{min}$ for anomalies in $\mathcal{J}_4$ (d) $a_{i,j}^{min}$ for anomalies in $\mathcal{J}_8$

Figure 6. Images of the minimum magnitude of the anomaly in Figure 5(a) to guarantee a $P_d = 0.95$ and $P_f = 0.05$ in binary hypothesis tests involving this anomaly structure and elements of $\mathcal{J}$. Note that while the scales in these images are different the magnitudes are all less than 0.6.

Note that if a_i is greater than $a_i^{max,min}$, a BHT with the anomaly $\bar{\gamma}_i$ given by $\mathcal{B}_i a_i$ will meet the $P_{d,i,j}$ and $P_{f,i,j}$ specification regardless of both the amplitude as well as the location of γ_j, i.e. the performance will be independent of j. On the other hand if a_i is less than $a_i^{min,min}$ then for *every* j there will be some range of amplitudes a_j for which the performance specifications will not be achieved. Now, from the first row of Table 2, we see that for an anomaly with geometric structure in Figure 5(a), an ABR of 2.11 ensures that any binary hypothesis test in which this structure is compared to a member of $\mathcal{J}$ will meet the performance specifications of $P_{d,i,j} = 0.95$ and $P_{f,i,j} = 0.05$. Alternatively, if the ABR falls below 0.56 then for all structures in $\mathcal{J}$, (i.e. all $\mathcal{B}_j$) the performance of the BHT will fail to meet the $P_{d,i,j}$ and $P_{f,i,j}$ requirements for some range of a_j. Similar results hold for the second anomaly structure located closer to the left side except that in this case, the required values of the ABR are smaller.

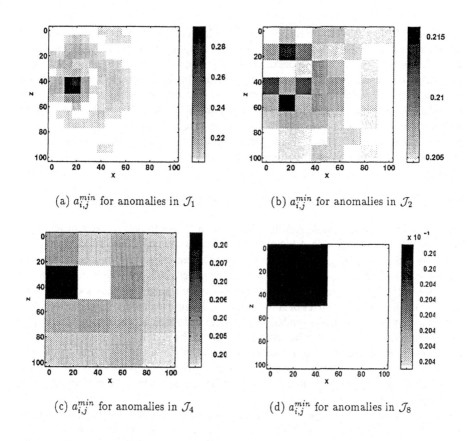

(a) $a_{i,j}^{min}$ for anomalies in $\mathcal{J}_1$ (b) $a_{i,j}^{min}$ for anomalies in $\mathcal{J}_2$

(c) $a_{i,j}^{min}$ for anomalies in $\mathcal{J}_4$ (d) $a_{i,j}^{min}$ for anomalies in $\mathcal{J}_8$

Figure 7. Images of the minimum magnitude of the anomaly in Figure 5(b) to guarantee a $P_d = 0.95$ and $P_f = 0.05$ in binary hypothesis tests involving this anomaly structure and elements of $\mathcal{J}$. Again, the scales in (a) through (d) are all different; however the overall range of values is between 0.2 and 0.3.

Table 2. Minimum and maximum anomaly-to-background ratio associated with the smallest and largest values for $a_{i,j}^{min}$ for the anomaly structures in Figure 5(a) and 5(b).

Anomaly $\bar{\gamma}_i$	Maximum ABR	Minimum ABR
Rightmost (Figure 5(a))	0.49	0.24
Leftmost (Figure 5(b))	2.11	0.56

6.   A Multiscale Algorithm for Anomaly Characterization

In this section we describe and analyze a multiscale, decision-theoretic algorithm to determine the positions, sizes and magnitudes of an unknown number of anomalous structures in region A. We begin with a small collection of relatively large rectangular areas in which anomalies *may* be located. Each region represents a top-level node in a tree of finer-scale subdivisions of A. We next use a decision-directed procedure for determining how best to move from one level of the tree, corresponding to a collection of coarse-scale hypotheses, to the next level in which anomalies are better localized using smaller-scale rectangles. The result of this procedure is a collection of rectangular areas of varying sizes and positions where we believe anomalies exist. To limit the number of targeted areas which contain no anomalies, the algorithm concludes with a pruning step where we also estimate the magnitudes of the final group of chosen anomaly structures.

6.1.   A Scale Recursive, Decision Driven Detection Algorithm

The first step in our detection algorithm involves an M-ary Hypothesis test in which we consider 10 ways to subdivide A in order to better localize anomalous structures. As seen in Figure 8 the first configuration corresponds to the presence of a coarse scale anomaly with support over all of A. This particular structure indicates that *no* further decomposition is warranted. The next four possibilities each allows for a single anomaly localized to the top, bottom, left and right halves of A respectively. Because anomalies might lie both in the left/right as well as the top/bottom halves, the sixth and seventh structures in Figure 8 are included. Since multiple anomalies may be present in the region, the eighth configuration corresponds to the presence of one anomaly located in the left half and one in the right while the ninth presents the analogous situation but for the top and the bottom. Finally, for this initial decomposition only, we consider the last case where we conjecture that *no* anomalous regions exists in A.

Given the 10 choices in Figure 8, we formulate a 10-ary hypothesis testing problem the solution of which is obtained using the Generalized Likelihood Ratio Test (GLRT) discussed in Section 3.2. Using (17) we compute the values of the generalized log-likelihood function for each of the hypotheses under consideration. From Figure 8, if H_0 is chosen, no further decomposition occurs and we conclude that there is a single anomaly covering the entire region of interest. If H_9 is selected, the algorithm terminates with the conclusion that there is no anomaly in region A. Otherwise, we decompose that hypothesis with the largest generalized log-likelihood value.

Our scale-recursive decomposition of A continues by essentially repeating the hypothesis testing procedure for each of the subregions indicated by the initial 10-ary hypothesis test as being of interest. For example, consider the case where H_3 is chosen. Referring to Figure 8, this selection corresponds to an anomaly located in the left half of A. In an effort to better localize the anomalous activity in this region, we consider an M-ary hypothesis test similar to that described in the previous paragraph but where the underlying area involved in the decomposition is now the left half of A rather than all of A. While the subdivision is of a rectangular region as opposed to a square area, the form of the hypotheses fundamentally remains the same as in those displayed in Figure 8 in that we consider the possibilities of anomalies located in the top, bottom, left, and right halves, etc. of this long and thin structure. We note that the first of these nine hypotheses, H_0, corresponds to the case where no further decomposition of the left half is warranted and thus serves as a means of

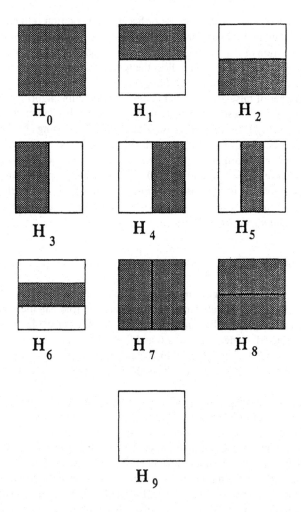

Figure 8. Geometric structures of the possible decompositions used at each stage of our decomposition of A. The darkly shaded regions indicate the areas where anomalous structures are hypothesized to exist. While the figure illustrates the decomposition of a square region, analogous subdivision schemes are used for rectangular areas as well with the fundamental idea being the presence of anomalies in the top, bottom, left, right, etc.

terminating the scale recursive search over this region of A. Instead of ten, there are only nine hypotheses as we no longer include the possibility that no anomaly exists in the left half of A since the previous iteration indicated that *somewhere* in the left side there exists an anomaly.

This nine-hypothesis GLRT is repeated recursively beginning with the regions selected in the initial decomposition of A. This decision-theoretic localization process continues until no further subdivision in a particular region is warranted based upon the selection of the

H_0 hypothesis at some stage of the process or because no addition refinement is possible because the structures under consideration are too small. Thus at the end of our scale-recursive decomposition of A we have a collection of rectangular regions where anomalous structures are likely to exist. We then collect the wavelet-domain representations of these rectangles as columns in a matrix labeled $\mathcal{B}_{leaf}$.

To limit the number of false alarms generated by our detection algorithm, we retain only those structures in $\mathcal{B}_{leaf}$ corresponding to sufficiently "detectable" anomalies. Specifically, we begin computing $\hat{a}_{leaf}$, the amplitudes associated with $\mathcal{B}_{leaf}$, using (14) with $\mathcal{B}_i$ replaced by $\mathcal{B}_{leaf}$. Next, for each column of $\mathcal{B}_{leaf}$, we calculate the minimum required amplitude to guarantee a set level of performance from a detectability-type hypothesis test developed in Section 4 (here we use $P_d = 0.80$ and $P_f = 0.10$). The final estimated anomaly structure generated by our algorithm is composed of those columns of $\mathcal{B}_{leaf}$ and elements of a_{leaf} corresponding to anomalies whose amplitudes exceed this required minimum and we label these estimates $\hat{B}$ and $\hat{a}$ respectively.

6.2. Algorithm Analysis

The scale-recursive detection algorithm described in Section 6.1 requires that we be able to identify successfully large-scale structures covering the true, smaller-scale anomalies. The results of the distinguishability analysis suggest that the correct large-scale structures are likely to be selected. Indeed, Figures 6 and 7 showed that the largest values of $a_{i,j}^{min}$ corresponded to those j in $\mathcal{J}$ which overlap anomaly i. From this, we conclude that small-scale anomalies "look" most like those large-scale counterparts located in the same region of A.

To further verify this intuition, we undertake a more detailed performance analysis of the GLRT used in the detection algorithm. Specifically, we consider the case where a single anomaly, $\bar{g}^*$, of unknown amplitude exists at some fine scale and we perform a generalized binary hypothesis in which the two hypotheses correspond to coarse scale structures one of which covers $\bar{g}^*$ and one of which does not. We are interested in examining how the probability of correctly choosing the overlapping structure (which we call the probability of detection for these experiments) using the GLRT of Section 3.2 varies with the scale and position of the non-overlapping alternate as well as the amplitude of the true anomaly. High detection probabilities reflect favorably on the GLRT-based approach of the scale-recursive algorithm.

Following the notation of (17), let $l_1(\eta)$ be the statistic associated with the overlapping anomaly hypothesis and $l_0(\eta)$ be the statistic for the non-overlapping case. From (15) and (16), the probability of choosing the overlapping structure given knowledge of $\bar{\gamma}^* = \mathcal{W}_g \bar{g}^*$ is

$$\text{Prob}[L_1(\eta) > 0 | \bar{\gamma}^*] = \text{Prob}[l_1(\eta) - l_0(\eta) > 0 | \bar{\gamma}^*]. \tag{29}$$

Upon substituting (14) into (16) and using (17), straightforward linear algebra demonstrates that the random variable $L_1(\eta)$ may be written as

$$L_1(\eta) = x_1^2(\eta) - x_0^2(\eta) \tag{30}$$

where the two-vector $x(\eta) = [x_1(\eta) \ x_0(\eta)]^T$ is

$$x(\eta) = \mathcal{B}_{10}^T \Theta^T P_\eta^{-1} \eta \sim \mathcal{N}(\mathcal{B}_{10}^T \Theta^T P_\eta^{-1} \Theta \bar{\gamma}^*, \mathcal{B}_{10}^T \Theta^T P_\eta^{-1} \Theta \mathcal{B}_{10}) \qquad (31)$$

and for $j = 0, 1$

$$\mathcal{B}_{10} = [s_1 \mathcal{B}_1 \ s_0 \mathcal{B}_0] \qquad (32a)$$

$$s_j^2 = \frac{1}{2}[P_j(1 + \alpha P_j)] \qquad (32b)$$

$$P_j = (\mathcal{B}_j^T \Theta^T P_\eta^{-1} \Theta \mathcal{B}_j + \alpha)^{-1} \qquad (32c)$$

From (30), $\mathrm{Prob}[L_1(\eta) > 0|\bar{\gamma}^*] = \mathrm{Prob}[\ |x_1(\eta)| > |x_0(\eta)|\ |\bar{\gamma}^*]$ which is the integral of the probability density function for $x(\eta)$ defined in (31) over the shaded region in Figure 9.

In Figure 10, detection probabilities are displayed for binary hypothesis tests where $\bar{g}^*$ is the structure in Figure 5(a) and the hypotheses are pairs of structures from $\mathcal{J}$. For example, the shade of dark region in the lower left corner in Figure 10(a) is $\mathrm{Prob}[L_1(\eta) > 0\ |\bar{\gamma}^*]$ for the BHT where the first hypothesis is the large structure overlapping the true, smaller size anomaly (represented by the white region in Figure 10(a)) and the alternate hypothesis is the 8×8 pixel lower left corner of A. Similar interpretations hold for the other two dark areas in Figure 10(a) and for each of the smaller square areas in Figures 10(b)–(c). For all of these images, the ABR for the true, small anomaly is set to 1.5. Figures 10(a)–(c) indicate the manner in which the detection performance of the GLRT-based algorithm depends upon the scale of the hypotheses relative to that of the true anomaly. At the coarsest scale, detection probabilities are about 60%. However, for all finer scales, P_d rises sharply with the lowest values confined to structures which are close to the true anomaly.

In Figure 10(d), we display the minimum P_d at each scale as a function of true anomaly's ABR. For example, the points on each of the three curves at an ABR of 1.5 are the minimum P_d values in each of the three images in Figure 10(a)–(c). From these curves we see that at the coarsest scale, even at high ABRs, the detection probabilities reach about 80%. As expected, when the hypotheses are drawn from the finer scales, the minimum P_d rises quickly to close to 100%.

The results in Figure 10 indicate that if the scale-recursive anomaly detection algorithm developed in Section 6.1 correctly identifies the coarse scale structures overlapping the true anomalies, then the detection performance at finer scales should be quite good even at ABRs less than 1. Also, because the *lowest* detection probabilities at fine scales are associated with structures close to the true structure, it is anticipated that the scale-recursive detection algorithm should be very successful in producing estimates of anomalies which are "sufficiently close" to the truth if not exactly the truth. This idea will be made more precise in Section 6.3.

The analysis in this section indicates that the primary difficulty associated with the algorithm is that coarse scale detection probabilities can be low. To overcome the potential problem of selecting the wrong area or areas of A for further refinement at coarse scales

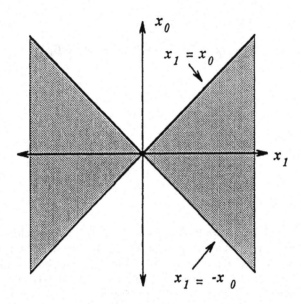

Figure 9. Integration region in $x_1 - x_0$ space for evaluation of $\text{Prob}[L_1(\eta) > 0 | \bar{\gamma}^*]$ in (29)

we modify the scale-recursive algorithm in the following manner. At the opening stage of the algorithm, rather than accepting the *single* hypothesis with the largest generalized log-likelihood value, we consider further refinement of A based upon those hypotheses corresponding to the *four* largest log-likelihood values (excluding H_0 and H_9). As will be seen in Section 6.3, despite the additional computational requirements of this approach, the overall complexity of the algorithm remains rather low. Finally, we note that one could extend this strategy of keeping additional structures for further refinement to more than just the first stage of the algorithm and could retain fewer or greater than four alternatives; however for the application of interest here, the choices described above were sufficient.

6.3.   *Examples*

In this section, we examine the performance of the scale-recursive algorithm described in Sections 6.1 and 6.2. First, we use Monte Carlo studies to verify the ability of this approach to detect anomalous structures. The quantities of interest here are the sample probability of detection, $\bar{P}_d$, the sample average value of the number of false alarms per pixel $\bar{P}_f$, and the sample probability of error, $\bar{P}_e$. We say that a particular rectangular anomaly, $\bar{\gamma}^*$, has been detected if there exists a column in $\hat{B}$ which is sufficiently close to $\bar{\gamma}^*$. Specifically, we define a "region of ambiguity" associated with the anomaly structure currently under investigation. This area is constructed such that anomaly structures identified in this region are "essentially indistinguishable" from the true anomaly. More formally, we compute the probability of

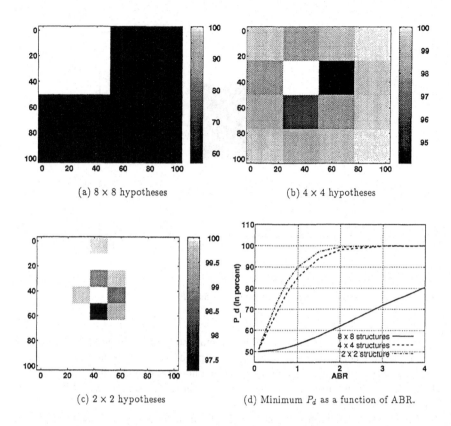

(a) 8×8 hypotheses

(b) 4×4 hypotheses

(c) 2×2 hypotheses

(d) Minimum P_d as a function of ABR.

Figure 10. In (a)–(c) detection probabilities are displayed for binary hypothesis tests where $\bar{g}^*$ is the structure in Figure 5(a) and the hypotheses are pairs of structures from $\mathcal{J}$. For each such test, one of the hypotheses is a larger scale structure overlapping $\bar{g}^*$ while the second structures is from the same scale as the first but is disjoint from the pixels of $\bar{g}^*$. The shade of each square in (a)–(c) is the probability of correctly choosing the overlapping structure when the alternate is the anomaly occupying the square under investigation. The ABR for the true structure is 1.5. The minimum P_d at each scale as a function of true anomaly's ABR is shown in (d).

successfully distinguishing $\bar{\gamma}^*$ from each member of $\mathcal{J}$ in a binary hypothesis test of the form in (28a)–(28b). For each such test, the amplitudes of the two anomalies are chosen so that relative to the anomaly-free background, the two structures are equally detectable (i.e. they individually have the same d^2 value as defined with $P_d = 0.85$ and $P_f = 0.10$ in (18) and (26).) A pixel in A is said to be in the ambiguity region if (1) there exists a member of $\mathcal{J}$ which is nonzero on that pixel and (2) the probability of distinguishing that element of $\mathcal{J}$ from $\bar{\gamma}^*$ is below a given threshold, taken as 0.85 for all problems considered in this section. Finally, for an estimated structure to be called a detection the area of intersection between it and the region of ambiguity must be at least a quarter of the area of the estimated structure. Such a definition implies a constraint on the localization of an estimated anomaly in both space and scale before we will call it a detection. As an example, the region of

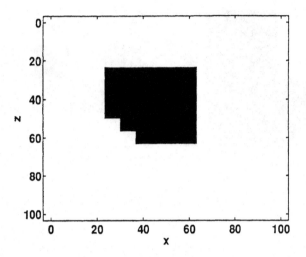

Figure 11. Region of ambiguity for structure shown in Figure 5 for $P_d = 0.85$.

ambiguity at $P_d = 0.85$ associated with the anomaly structure in Figure 5 is displayed in Figure 11. The elements of $\hat{B}$ which do not correspond to detections are taken to be false alarms and the per-pixel false alarm rate, $\bar{P}_f$, is defined as the total number of false alarm pixels divided by the number of pixels in region A. Finally, the sample probability of error is $\bar{P}_e = 1 - \bar{P}_d + \bar{P}_f$.

We also examine the computational complexity of the scale-recursive algorithm. The complexity of the algorithm is quantified in terms of the number of Generalized Likelihood Ratio Tests (GLRTs) which must be performed in the processing of the data. As the spatial decomposition of region A is driven by the noisy data, the number of GLRTs will vary from one data set to the next. Thus, for a particular $\bar{g}^*$, the computational performance is based upon the average number of required GLRTs required per iteration of the corresponding Monte-Carlo.

Finally as discussed in Section 1, the detection algorithm results are used to improve the solution to the full reconstruction inverse problem. From our model for γ in (6), the estimate of the overall conductivity is the sum of the estimates of $\bar{\gamma}$ and $\tilde{\gamma}$, denoted $\hat{\bar{\gamma}}$ and $\hat{\tilde{\gamma}}$ respectively, where $\hat{\tilde{\gamma}} = \hat{B}\hat{a}$ is provided by our scale-recursive detection algorithm. Now, the linear least-squares estimate (LLSE) of $\bar{\gamma}$ developed in [37, 40] is based upon the assumption that *no* anomalies exist in the data; however, the output of the detection algorithm provides additional information through $\hat{\tilde{\gamma}}$ as to the structure of the conductivity field. To make use of the information in order to improve the estimate $\hat{\bar{\gamma}}$, we define $\hat{\bar{\gamma}}_c$ as the LLSE of $\bar{\gamma}$ based upon a "corrected" data set in which the effects of $\hat{\tilde{\gamma}}$ have been removed. Mathematically this corrected estimate takes the form

$$\hat{\bar{\gamma}}_c = P\Theta^T R^{-1}[\eta - \Theta\hat{B}\hat{a}] \tag{33}$$

where $P = (\Theta^T R^{-1} \Theta + P_0^{-1})^{-1}$ is the error covariance matrix for nominal LLSE. Thus, the estimate of the overall conductivity field is

$$\hat{\gamma} = \hat{\bar{\gamma}}_c + \hat{\tilde{\gamma}} = P\Theta^T R^{-1} \eta + [I - P\Theta^T R^{-1}\Theta]\hat{B}\hat{a} \tag{34}$$

where we recognize the term $P\Theta^T R^{-1}\eta$ as the uncorrected LLSE estimate [49].

Unless otherwise stated, the data upon which the examples are based are generated using the Born-based measurements model in (2) for the scattering experiments described in Table 1. For all cases consider, the background conductivity, g_0, is set to 1 S/m and at the highest ABRs of interest, the anomaly amplitudes are only 0.7 S/m. As discussed in [25], under these circumstances the Born approximation is known to be valid. In Section 7, we discuss issues associated with extending the work in this paper to account for the underlying non-linearity associated with the inverse conductivity problem. Finally, for all experiments the parameter α in (14) is set to 0.25.

6.3.1. The Single Anomaly Case

We begin by considering the case where it is known that there is a single anomaly of unknown amplitude and location in region A. Given that there is only one structure, the combinatorial complexity associated with an "exhaustive search" for the anomaly is sufficiently low that we shall compare both the detection/false-alarm performance as well as the complexity of the scale-recursive approach against an alternate algorithm akin to a multi-scale matched filter. This algorithm detects the single anomaly by computing the GLRT for each of the structures in family $\mathcal{J}$ taking that element of $\mathcal{J}$ associated with the largest GLRT statistic as the estimate. Because this method is multiscale in nature and has a fixed number of GLRTs per Monte-Carlo iteration (since there are a fixed number of structures in $\mathcal{J}$) it allows for a fair comparison against which we can judge the performance of the scale-recursive algorithm. For the scale-recursive method, we shall account for the knowledge that there is only a single anomaly in A by retaining only the column of $\mathcal{B}_{leaf}$ associated with the most likely anomaly structure. Finally, for this example, the true anomaly structure is shown in Figure 5 and the SNR for all scattering experiments is 10.

In Figure 12 we show $\bar{P}_d$, $\bar{P}_f$, $\bar{P}_e$ and the average number of GLRTs per Monte-Carlo iteration as a function of anomaly-to-background ratio obtained after 500 Monte-Carlo iterations. The solid lines are the results for scale-recursive algorithm and the dashed lines indicate the performance of the multi-scale exhaustive search procedure. Figure 12(a) indicates that at low ABRs, the scale-recursive approach tends to have a higher detection probability than the exhaustive search with a slightly higher probability of false alarm. Even for the low ABR of 0.50, $\bar{P}_d$ is well above 50% and rises to above 90% for ABR values greater than one. At high ABRs the performance of the two algorithms is about the same. Despite the slightly higher $\bar{P}_f$ of the scale-recursive approach, the overall error probability is lower for the scale-recursive method at these small ABRs. Finally, from 12(c) the computational complexity of the scale-recursive characterization algorithm is seen to be roughly constant across the ABR range at 65% that of the exhaustive search.

In Figure 13(a) we display one realization of $g = \bar{g} + \tilde{g}$ obtained in our Monte Carlo process at an ABR of 1.5. Using the LLSE to perform the full reconstruction as in [40]

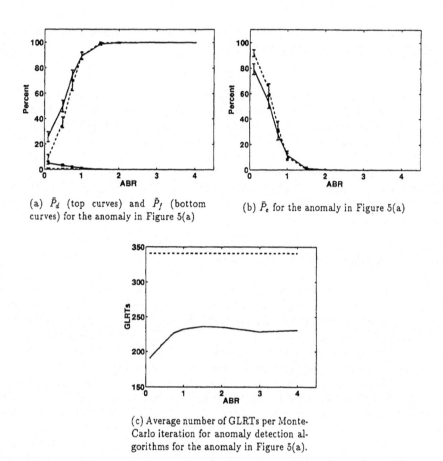

(a) $\bar{P}_d$ (top curves) and $\bar{P}_f$ (bottom curves) for the anomaly in Figure 5(a)

(b) $\bar{P}_e$ for the anomaly in Figure 5(a)

(c) Average number of GLRTs per Monte-Carlo iteration for anomaly detection algorithms for the anomaly in Figure 5(a).

Figure 12. Performance curves as a function of ABR obtained after 500 Monte-Carlo iterations for the anomaly in Figure 5(a). Solid lines = results for scale-recursive algorithm. Dashed lines = results for multi-scale exhaustive search. The error bars are drawn at the plus/minus two standard deviation level.

results in the image in Figure 13(b). By incorporating the results of the scale-recursive detection algorithm into the inversion procedure through the use of (34), we obtain the estimate of the overall conductivity field shown in 13(c). Thus, successful identification of the highly parameterized anomaly structures can significantly improve localization both in space and scale and the GLRT procedure results in an accurate estimate of the structure's amplitude. Also, the details in the remainder of the estimate do in fact reflect the coarse scale, fractal features of the conductivity profile in Figure 13(a).

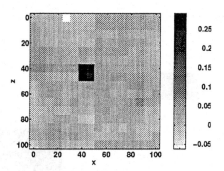

(a) Anomaly in fractal background (ABR=1.5)

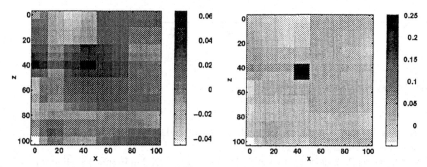

(b) Reconstruction of conductivity in (a) using LLSE

(c) Reconstruction of conductivity in (a) using (34)

Figure 13. Comparison of reconstructed conductivity profile using the LLSE of [40] and an estimate based upon the output of the scale-recursive anomaly detection algorithm. The true conductivity is shown in (a) and contains a single anomaly near the center of the region. The LLSE is shown in (b) and the estimate obtained from (34) is illustrated in (c). Here we see that the use of the information from the detection algorithm allows for the successful localization of the anomaly in space and scale without sacrificing our ability to resolve the fractal features of the conductivity profile in (a). Additionally, the GLRT procedure results in an accurate estimate of the anomaly's amplitude.

ι

6.3.2. *The Multiple Anomaly Case*

We now turn our attention to the case where multiple anomalies exist in region A.[4] Lifting the single anomaly assumption causes the computational complexity of an exhaustive-search-type of approach to be prohibitive in that one would be required to examine the likelihood of all combinations of all non-overlapping, structures in a collection such as $\mathcal{J}$ assuming separately $n = 1$ then $n = 2$ through $n = N_{max}$ anomalies exist in region A where N_{max} is a pre-determined maximum number of anomalies. Thus, here we present only the results

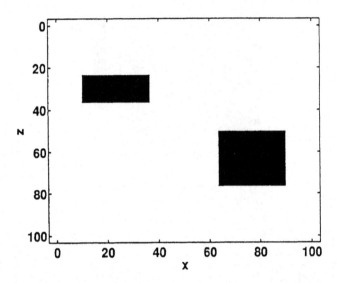

Figure 14. Two-region anomaly structure

of the scale-recursive detection algorithm. In particular, we explore the performance for the anomaly configuration in Figure 14.

The Monte-Carlo results for this experiment are displayed in Figure 15 where the top two curves of (a) correspond to the individual $\bar{P}_d$ statistics for the two anomalies and the lowest of the three curves is a plot of $\bar{P}_f$. Here we see that both structures are quite easily detected with a $\bar{P}_d$ of well over 90% even at the low ABR of one. As is expected, removing the single-anomaly assumption causes the algorithm to retain a greater number of candidate structures (including the true anomalies) thereby raising $\bar{P}_f$ above that seen in Section 6.3.1.

In Figure 15(c) we plot the average number of GLRTs as a function of ABR. Note that at *worst* the complexity of this algorithm is still well below the complexity of the single-anomaly exhaustive search algorithm and only about 30% greater than the complexity of the single-anomaly scale-recursive algorithm. Thus, despite the fact that the multiple anomaly problem is, from a combinatorial viewpoint, significantly more complex than the single anomaly case, we see that the scale-recursive localization method represents a highly efficient and accurate means of localizing an unknown number of structures in the region of interest.

In Figure 16, we compare the full reconstruction results obtained from the LLSE to those where (34) is used to estimate the underlying conductivity for one run of the Monte-Carlo at an ABR of 1.5. From Figure 16(b) we see that the LLSE is successful in reconstructing the structure on the left; however, the lower amplitude/more pixel anomaly is almost completely undetected. Figure 16(c) indicates that the incorporation of the information from the anomaly detection algorithm significantly improves the localization in space as well as

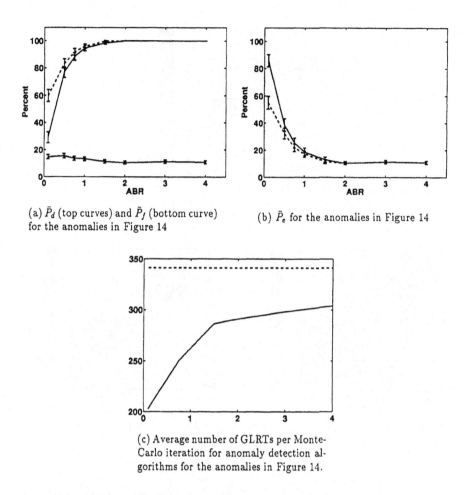

(a) $\bar{P}_d$ (top curves) and $\bar{P}_f$ (bottom curve) for the anomalies in Figure 14

(b) $\bar{P}_e$ for the anomalies in Figure 14

(c) Average number of GLRTs per Monte-Carlo iteration for anomaly detection algorithms for the anomalies in Figure 14.

Figure 15. Performance curves obtained after 500 Monte-Carlo iterations of scale-recursive detection algorithm for the anomalies in Figure 14. Solid lines in (a) and (b) are detection and error probabilities for the upper left anomaly while dashed lines are for lower right anomaly. The error bars are drawn at the plus/minus two standard deviation level. In (c), the computational complexity associated with this scenario is shown by the solid line. For comparison, the dashed line is the complexity associated with the single anomaly exhaustive search.

scale of both anomaly structures, especially the rightmost. Finally, the anomaly amplitudes are better estimated using the GLRT method.

7. Conclusion and Future Work

In this paper, we have presented a framework based upon techniques from the areas of multiscale modeling, wavelet transforms, and statistical decision and estimation theory for

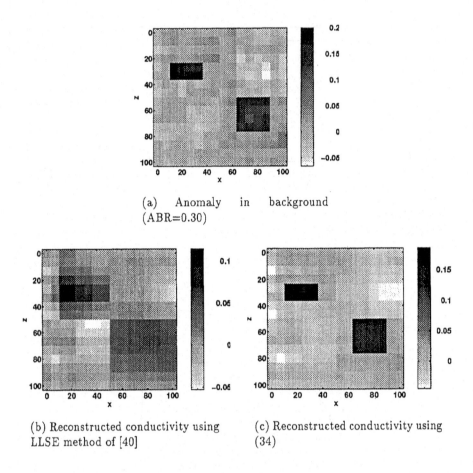

(a) Anomaly   in   background
(ABR=0.30)

(b) Reconstructed conductivity using   (c) Reconstructed conductivity using
LLSE method of [40]   (34)

Figure 16. Comparison of reconstructed conductivity profile using the LLSE of [40] and an estimate based upon the output of the scale-recursive anomaly detection algorithm. The true conductivity is shown in (a) and contains a two anomalies. The LLSE is shown in (b) and the estimate obtained from (34) is illustrated in (c). Here we see that the use of the detection information allows for the successful localization of both anomaly structures and offers a significant improvement over the LLSE in localizing the anomaly in the lower right.

addressing a variety of issues arising in anomaly detection problems. Beginning with a linear model relating the data and the quantity to be reconstructed, we use the wavelet transform to take the problem from physical space to scale space where computational complexity is reduced for a wide variety of problems [1, 4, 41] and where we are able to take advantage of the rich and useful class of models recently developed for describing the structure of the medium in the absence of anomalous activity [21, 35, 47, 50]. The problems of characterizing the number, positions, and magnitudes of anomaly structures was formulated using the tools of statistical decision theory. To understand how the physics of the problem and the constraints on the geometry of the data collection process affect our ability to isolate anomalous regions, we defined and explored the issues of anomaly detectability and distinguishability. This analysis led to the development of a scale-recursive algorithm employing a sequence of Generalized Likelihood Ratio Tests for extracting anomaly information from data.

This work was presented in the context of a linearized inverse scattering problem arising in geophysical prospecting. The same scattering model is encountered in a variety of other fields where some form of energy is used to probe a lossy medium [18–20, 22, 29, 31]. More generally, the analysis and algorithmic methods developed in this work require only a measurements model of the form in (2) and are thus relevant for any linear inverse problem (e.g. computed tomography) in which anomaly characterization is of interest.

An important extension of the work presented here is in development of algorithms and analysis methods for detecting anomaly structures using the *nonlinear* physics governing the relationship between the conductivity and the observed scattered electric field. The primary difficulty here is maintaining or improving the detection/false-alarm performance of the current method while retaining the low computation complexity in an algorithm based upon a significantly more complex scattering model. In [38] we present preliminary results for one form of a scale-recursive anomaly characterization algorithm using the computationally efficient, nonlinear Extended Born Approximation [48]. Further work remains in the exploration of detectability and distinguishability in the nonlinear context and the extension of this approach to higher frequency (e.g. microwave) problems.

Another avenue of research is in the use of alternate methods for progressively dividing the region of interest. The problem of anomaly detection is similar to that of image segmentation in that the goal of both is to partition a two-dimensional grid of pixels into disjoint regions. The primary difference between these two problems is the data. In the segmentation case the data are the pixels in the image whereas we wish to do the anomaly localization given the significantly less informative observations of scattered radiation. For the segmentation problem, significant work has been performed in the use of hierarchical methods for performing this decomposition. For example, segmentation techniques have been developed where (a) small structures are merged into larger regions [36, 42] and (b) both splitting as well as merging operations are used in the segmentation process [28]. Examining the utility of merge- and split/merge-algorithms for the anomaly detection problem would be of considerable interest especially as a means of overcoming the difficulties of detecting small-scale structures using large scale hypotheses.

Acknowledgements

This work was supported in part by the Air Force Office of Scientific Research under Grant AFOSR-92-J-0002, and the Advanced Research Project Agency under Air Force Grant F49620-93-1-0604 and the Office of Naval Research under Grant N00014-91-J-1004. The work of the first author was also supported in part by a US Air Force Laboratory Graduate Fellowship.

Notes

1. The notation $x \sim \mathcal{N}(m, P)$ indicates that the random vector x has a Gaussian probability distribution with mean m and covariance matrix P.
2. Note that in the contexts where the binary testing scenario is to be explored, the values of a_0 and a_1 are assumed known so that a generalized test is not required.
3. For illustrative purposes only, in Figure 3 it is assumed that the major axis of the ellipse is oriented at an angle less than 90° from the a_0 axis. While this is not necessarily the case, the analysis which follows is independent of which axis is the major and which the minor.
4. Note that in this multi-anomaly case, the ABR is used to determine the magnitude of each structure individually. For example at an ABR of one, the amplitude of the left anomaly is set so that if it were the only structure in the medium, the ABR would be one.

References

1. B. Alpert, G. Beylkin, R. Coifman, and V. Rokhlin, "Wavelets for the fast solution of second-kind integral equations," *SIAM J. on Scient. Comput.*, vol. 14, no. 1, 1993, pp. 159–184.

2. M. Bertero, C. De Mol, and E. R. Pike, "Linear inverse problems with discrete data. I: General formulation and singular system analysis," *Inverse Problems*, vol. 1, 1985, pp. 301–330.

3. M. Bertero, C. De Mol, and E. R. Pike, "Linear inverse problems with discrete data. II: Stability and regularisation," *Inverse Problems*, vol. 4, 1988, pp. 573–594.

4. G. Beylkin, R. Coifman, and V. Rokhlin, "Fast wavelet transforms and numerical algorithms I," *Communications on Pure and Applied Mathematics*, vol. 44, 1991, pp. 141–183.

5. Y. Bresler, J. A. Fessler, and A. Macovski, "A Bayesian approach to reconstruction from incomplete projections of a multiple object 3D domain," *IEEE Trans. on Pattern Analysis and Machine Intelligence*, vol. 11, no. 8, 1989, pp. 840–858.

6. K. E. Brewer and S. W. Wheatcraft, "Including multi-scale information in the characterization of hydraulic conductivity distributions," In E. Foufoula-Georgiou and P. Kumar (eds.), *Wavelets in Geophysics*, vol. 4 of *Wavelet Analysis and its Applications*, pp. 213–248. Academic Press, 1994.

7. S. R. Brown, "Transport of fluid and electric current through a single fracture," *Journal of Geophysical Research*, vol. 94, no. B7, 1989, pp. 9429–9438.

8. W. C. Chew and Y. M. Wang, "Reconstruction of two-dimensional permittivity distribution using the distorted born iterative method," *IEEE Trans. Medical Imaging*, vol. 9, no. 2, pp. 218–225, 1990.

9. W. C. Chew, *Waves and Fields in Inhomogeneous Media*, New York: Van Nostrand Reinhold, 1990.

10. K. C. Chou, S. A. Golden, and A. S. Willsky, "Multiresolution stochastic models, data fusion and wavelet transforms," Technical Report LIDS-P-2110, MIT Laboratory for Information and Decision Systems, 1992.

11. K. C. Chou, A. S. Willsky, and R. Nikoukhah, "Multiscale recursive estimation, data fusion, and regularization," *IEEE Trans. Automatic Control*, vol. 39, no. 3, 1994, pp. 464–478.

12. K. C. Chou, A. S. Willsky, and R. Nikoukhah, "Multiscale systems, Kalman filters, and Riccati equations," *IEEE Trans. Automatic Control*, vol. 39, no. 3, 1994, pp. 479–492.

13. D. J. Crossley and O. G. Jensen, "Fractal velocity models in refraction seismology," In C. H. Scholtz and B. B. Mandelbrot (eds.), *Fractals in Geophysics*, pp. 61–76. Birkhauser, 1989.

14. I. Daubechies, "Orthonormal bases of compactly supported wavelets," *Communications on Pure and Applied Mathematics*, vol. 41, 1988, pp. 909–996.

15. A. J. Devaney, "Geophysical diffraction tomography," *IEEE Trans. on Geoscience and Remote Sensing*, vol. GE-22, no. 1, 1984, pp. 3–13.

16. A. J. Devaney and G. A. Tsihrintzis, "Maximum likelihood estimation of object location in diffraction tomography," *IEEE Trans. ASSP*, vol. 39, no. 3, 1991, pp. 672–682.

17. A. J. Devaney and G. A. Tsihrintzis, "Maximum likelihood estimation of object location in diffraction tomography, part II: Strongly scattering objects," *IEEE Trans. ASSP*, vol. 39, no. 6, 1991, pp. 1466–1470.

18. D. C. Dobson, "Estimates on resolution and stabilization for the linearized inverse conductivity problem," *Inverse Problems*, vol. 8, 1992, pp. 71–81.

19. D. C. Dobson and F. Santosa, "An image-enhancement technique for electrical impedance tomography," *Inverse Problems*, vol. 10, 1994, pp. 317–334.

20. D. C. Dobson and F. Santosa, "Resolution and stability analysis of an inverse problem in electrical impedance tomography: Dependence on the input current patterns," *SIAM J. Appl. Math.*, vol. 54, no. 6, pp. 1542–1560.

21. P. Flandrin, "Wavelet analysis and synthesis of fractional Brownian motion," *IEEE Trans. Information Theory*, vol. 38, no. 2, pp. 910–917.

22. D. G. Gisser, D. Isaacson, and J. C. Newell, "Electric current computed tomography and eigenvalues," *SIAM J. Appl. Math.*, vol. 50, no. 6, pp. 1623–1634.

23. T. M. Habashy, W. C. Chew, and E. Y. Chow, "Simultaneous reconstruction of permittivity and conductivity profiles in a radially inhomogeneous slab," *Radio Science*, vol. 21, no. 4, pp. 635–645.

24. T. M. Habashy, E. Y. Chow, and D. G. Dudley, "Profile inversion using the renormalized source-type integral equation approach," *IEEE Transactions on Antennas and Propagation*, vol. 38, no. 5, pp. 668–682.

25. T. M. Habashy, R. W. Groom, and B. R. Spies, "Beyond the Born and Rytov approximations: A nonlinear approach to electromagnetic scattering," *Journal of Geophysical Research*, vol. 98, no. B2, pp. 1759–1775.

26. R. F. Harrington, *Field Computations by Moment Methods*, Macmillan Publ. Co., 1968.

27. J. H. Hippler, H. Ermert, and L. von Bernus, "Broadband holography applied to eddy current imaging using signals with multiplied phases," *Journal of Nondestructive Evaluation*, vol. 12, no. 3, pp. 153–162.

28. S. L. Horowitz and T. Pavlidis, "Picture segmentation by a tree traversal algorithm," *Journal of the ACM*, vol. 23, no. 2, 1976, pp. 368–388.

29. D. Isaacson, "Distinguishability of conductivities by electrical current computed tomography," *IEEE Trans. on Medical Imaging*, vol. MI-5, no. 2, 1986, pp. 91–95.

30. D. Isaacson and M. Cheney, "Current problems in impedance imaging," In D. Colton, R. Ewing, and W. Rundell (eds.), *Inverse Problems in Partial Differential Equations*, Ch. 9, pp. 141–149. SIAM, 1990.

31. D. Isaacson and M. Cheney, "Effects of measurement precision and finite numbers of electrodes on linear impedance imaging algorithms," *SIAM J. Appl. Math.*, vol. 51, no. 6, 1991, 1705–1731.

32. D. L. Jaggard, "On fractal electrodynamics," In H. N. Kritikos and D. L. Jaggard (eds.), *Recent Advances in Electromagnetic Theory*, pp. 183–224. Springer-Verlag, 1990.

33. J. M. Lees and P. E. Malin, "Tomographic images of *p* wave velocity variation at Parkfield, California," *Journal of Geophysical Research*, vol. 95, no. B13, 1990, pp. 21,793–21,804.

34. V. Liepa, F. Santosa, and M. Vogelius, "Crack determination from boundary measurements—Reconstruction using experimental data," *Journal of Nondestructive Evaluation*, vol. 12, no. 3, 1993, pp. 163–174.

35. S. G. Mallat, "A theory of multiresolution signal decomposition: The wavelet representation," *IEEE Trans. PAMI*, vol. 11, no. 7, 1989, pp. 674–693.

36. J. M. Beaulieu and M. Goldberg, "Hierarchy in picture segmentation: A stepwise optimization approach," *IEEE Trans. Pattern Analysis and Machine Intelligence*, vol. 11, no. 2, 1989, pp. 150–163.

37. E. L. Miller, "The application of multiscale and statistical techniques to the solution of inverse problems," Technical Report LIDS-TH-2258, MIT Laboratory for Information and Decision Systems, Cambridge, 1994.

38. E. L. Miller, "A scale-recursive, statistically-based method for anomaly characterization in images based upon observations of scattered radiation," In *1995 IEEE International Conference on Image Processing*, 1995.

39. E. L. Miller and A. S. Willsky, "A multiscale approach to sensor fusion and the solution of linear inverse problems," *Applied and Computational Harmonic Analysis*, vol. 2, 1995, pp. 127–147.

40. E. L. Miller and A. S. Willsky, "Multiscale, statistically-based inversion scheme for the linearized inverse scattering problem," *IEEE Trans. on Geoscience and Remote Sensing*, March 1996, vol. 36, no. 2, pp. 346–357.

41. E. L. Miller and A. S. Willsky, "Wavelet-based, stochastic inverse scattering methods using the extended Born approximation," In *Progress in Electromagnetics Research Symposium*, 1995. Seattle, Washington.

42. J. Le Moigne and J. C. Tilton, "Refining image segmentation by integration of edge and region data," *IEEE Trans. on Geoscience and Remote Sensing*, vol. 33, no. 3, 1995, pp. 605–615.

43. J. E. Molyneux and A. Witten, "Impedance tomography: imaging algorithms for geophysical applications," *Inverse Problems*, vol. 10, 1994, pp. 655–667.

44. D. J. Rossi and A. S. Willsky, "Reconstruction from projections based on detection and estimation of objects–parts I and II: Performance analysis and robustness analysis," *IEEE Trans. on ASSP*, vol. ASSP-32, no. 4, 1984, pp. 886–906.

45. K. Sauer, J. Sachs, Jr., and C. Klifa, "Bayesian estimation of 3D objects from few radiographs," *IEEE Trans. Nuclear Science*, vol. 41, no. 5, 1994, pp. 1780–1790.

46. A. Schatzberg, A. J. Devaney, and A. J. Witten, "Estimating target location from scattered field data," *Signal Processing*, vol. 40, 1994, pp. 227–237.

47. A. H. Tewfick and M. Kim, "Correlation structure of the discrete wavelet coefficients of fractional Brownian motion," *IEEE Trans. Information Theory*, vol. 38, no. 2, pp. 904–909.

48. C. Torres-Verdín and T. M. Habashy, "Rapid 2.5-D forward modeling and inversion via a new nonlinear scattering approximation," *Radio Science*, 1994, pp. 1051–1079.

49. H. L. Van Trees, *Detection, Estimation and Modulation Theory: Part I*. New York: John Wiley and Sons, 1968.

50. G. W. Wornell, "A Karhuenen-Loeve-like expansion for $1/f$ processes via wavelets," *IEEE Transactions on Information Theory*, vol. 36, 1990, pp. 859–861.

Multidimensional Systems and Signal Processing, 8, 185–217 (1997)

On the Scalability of 2-D Discrete Wavelet Transform Algorithms

JOSÉ FRIDMAN AND ELIAS S. MANOLAKOS   jfridman@cdsp.neu.edu, elias@cdsp.neu.edu
Communications and Digital Signal Processing (CDSP), Center for Research and Graduate Studies, Electrical and Computer Engineering Department, 409 Dana Research Building, Northeastern University, Boston, MA 02115

Abstract. The ability of a parallel algorithm to make efficient use of increasing computational resources is known as its *scalability*. In this paper, we develop four parallel algorithms for the 2-dimensional Discrete Wavelet Transform algorithm (2-D DWT), and derive their scalability properties on Mesh and Hypercube interconnection networks. We consider two versions of the 2-D DWT algorithm, known as the *Standard* (S) and *Non-standard* (NS) forms, mapped onto P processors under two data partitioning schemes, namely *checkerboard* (CP) and *stripped* (SP) partitioning. The two checkerboard partitioned algorithms on the *cut-through-routed* (CT-routed) Mesh are scalable as $M^2 = \Omega(P \log P)$ (Non-standard form, NS-CP), and as $M^2 = \Omega(P \log^2 P)$ (Standard form, S-CP); while on the *store-and-forward-routed* (SF-routed) Mesh and Hypercube they are scalable as $M^2 = \Omega(P^{\frac{3}{3-\gamma}})$ (NS-CP), and as $M^2 = \Omega(P^{\frac{2}{2-\gamma}})$ (S-CP), respectively, where M^2 is the number of elements in the input matrix, and $\gamma \in (0, 1)$ is a parameter relating M to the number of desired octaves J as $J = \lceil \gamma \log M \rceil$. On the CT-routed Hypercube, scalability of the NS-form algorithms shows similar behavior as on the CT-routed Mesh. The Standard form algorithm with stripped partitioning (S-SP) is scalable on the CT-routed Hypercube as $M^2 = \Omega(P^2)$, and it is unscalable on the CT-routed Mesh. Although asymptotically the stripped partitioned algorithm S-SP on the CT-routed Hypercube would appear to be inferior to its checkerboard counterpart S-CP, detailed analysis based on the proportionality constants of the isoefficiency function shows that S-SP is actually *more* efficient than S-CP over a realistic range of machine and problem sizes. A milder form of this result holds on the CT- and SF-routed Mesh, where S-SP would, asymptotically, appear to be altogether unscalable.

Key Words: wavelets, scalability, parallel processing

1. Introduction

The Continuous Wavelet Transform (CWT) is a time-scale decomposition representing the frequency content of finite energy signals as it evolves in time, analogously to the Short Time Fourier Transform (STFT) and the Gabor Transform (GT). The distinguishing property of the CWT is that its basis functions are shifts and translates of a single time- and frequency-localized function called *mother wavelet*. The CWT and the theory of *multiresolution analysis* have received a great deal of attention over the last ten years and have recently emerged as powerful signal processing frameworks [1], [2]. The Discrete Wavelet Transform (DWT) is the discrete-time discrete-scale counterpart of the CWT [3], [4], [5]. The DWT has been successfully applied to numerous and diverse application domains. Among them we mention data compression and transmission [6], [7], as an alternative to traditional time-scale representations (STFT and GT) using polynomial spline approximations [8], as an operator for fast numerical analysis [9], [10], [11], [12], and in image processing applications such as edge detection [13] and image coding [14].

Although computationally the 2-D DWT is faster than alternative time-frequency decompositions (a separable 2-D DWT on an $M \times M$ input matrix with length L filters is in $O(M^2 L)$, whereas the 2-D DFT is in $O(M^2 \log M)$), there are currently many application domains where the cost of a transform on M^2 is prohibitively high. One way to overcome this inherent limitation on computational cost is by means of parallel processing. Consider for instance a typical remote sensing satellite data stream, requiring one to two *weeks* of off-line computer time [15]. Each thematic scene from the Landsat satellite consists of nearly 300 megabytes, and every two weeks 20,000 images are transmitted to earth. Parallel processing provides the only viable alternative for reducing the computation time for obtaining a real-time response. A similar high-computational cost problem is also faced by digital industrial and consumer video systems in multimedia environments, dealing with compression and multi-rate transmission.

A few parallel algorithms and implementations of the 2-D DWT have been reported, see for example [16], [17]. However, to the best of our knowledge, scalability properties of the 2-D DWT algorithm have not been addressed to date. We have successfully applied a methodology for systematically synthesizing VLSI-amenable parallel arrays for computing the 1-D DWT [18], [19], [20], [21], known as *data dependence* analysis. In this paper we use a simple form of this data dependence analysis in the derivation of our 2-D DWT algorithms' performance; our findings on the scalability of a parallel 1-D DWT algorithm, along with experimental data on the MasPar MP-1, may be found in [22].

One of the main difficulties in parallel algorithm design arises from the greater multiplicity of possible solutions in the parallel domain, in contrast to single-processor architectures. A central design issue is *data partitioning*: the initial distribution and organization of input data among processors. In numerical algorithms like the 2-D DWT, the most common data partitioning schemes used are *checkerboard* and *stripped* partitioning, and variants thereof. The entire communication mechanism is heavily dependent on the initial data distribution in parallel algorithms, consequently determining system performance.

Three alternatives for *parallel system*[1] performance analysis are: (i) by deriving analytic expressions estimating the parallel execution time assuming a model for communication; (ii) by simulation using an event-driven software prototype; and (iii) by measuring actual experimental execution times on a specific architecture. Each method has advantages and limitations, and neither one can fully describe a particular system. Yet, systems which to some degree admit a performance modeling-type approach can be understood and analyzed at a deeper level, because carefully-devised modeling—accounting for its limitations—can offer a powerful and general descriptive tool to an extent that simulation and experimentation cannot. In this paper, we derive execution times of our four parallel 2-D DWT algorithms by performance modeling. Despite arguments over the limitations of modeling, it is nonetheless instrumental in revealing the fundamental mechanisms governing the behavior of the algorithms and their interrelationships.

We develop four 2-D DWT parallel algorithms and report on the results of a detailed scalability analysis based on the *isoefficiency* metric [23]. Our algorithms are developed on two widely-used processor networks: the Mesh and the Hypercube, and with two communication routing models: *virtual cut-through* (CT) and *store-and-forward* (SF) routing. We elaborate mostly on development and model details of the Mesh CT-routed algorithms because the 2-D DWT naturally maps into this topology, and also because the Mesh has recently received considerable attention and is widely regarded as cost-effective and highly scalable; commercially available Mesh computers include Intel's Paragon, MasPar's MP-1 and 2, ICE's MeshSP Superstation, and Parsytec's PowerPlus. The Hypercube versions of our algorithms arise as extensions to the Mesh algorithms; commercial machines based on the Hypercube include the nCUBE 2, Thinking Machine's CM-2, and Intel's iPSC/2.

We consider two variants of the 2-D DWT that, borrowing from Beylkin's terminology [9], we refer to as *Standard* and *Non-standard* forms. The application domain of the Non-standard 2-D DWT form (usually called Mallat's decomposition [4]) has been mostly in image processing [6], [7], while the Standard form has been mostly used as a fast linear operator in numerical analysis [9], [10], [11], [12], and has shown slight performance improvement in still image compression as compared to the NS-form [24]. In addition, the algorithms we describe are distinguished on the basis of data partitioning, being either of (row) *Stripped* or of *Checkerboard* type [23]. Thus, four different forms arise:

- Standard DWT with Stripped partitioning (**S-SP**),

- Standard DWT with Checkerboard partitioning (**S-CP**),

- Non-standard DWT with Stripped partitioning (**NS-SP**), and

- Non-standard DWT with Checkerboard partitioning (**NS-CP**).

Among these four versions of the 2-D DWT, S-SP may be implemented in two distinct phases: one consisting of purely local computations (to each processor) with no communication, and the other consisting of a global communication operation where every processor exchanges a distinct message with every other processor (known as an all-to-all personalized [23]). The difference between S-SP and the other three algorithms highlights the distinction between two parallel programming styles. One relies on communication primitives tuned for a specific native machine, like the PICL, ComPaSS and BLACS libraries [25], [26], having advantages for programmers as well as for automating the compilation process, and providing an attractive alternative for portability. On the other hand, the details involved in an algorithm's communication requirements may be left to the programmer (or compiler) in the form of a *message-passing* interface, which in some cases allow higher performance levels at the cost of increased programming complexity.

As one of the main results in this paper, we show that a 2-D DWT parallel algorithm based on a single global communication operation can, for a given range of architecture and problem-size parameters, exhibit comparable (and in some cases superior) performance relatively to an algorithm based on interleaving a number of local communication and computation stages. This result holds interest mostly in that, from the point of view of a purely asymptotic scalability analysis, the latter algorithms would appear to be superior (in the isoefficiency sense) than the former one based on a single global communication operation. In fact, the algorithm with a single global communication (when implemented on the Mesh) would appear to be altogether *un*scalable if evaluated through asymptotic isoefficiency metrics alone. In other words, we are establishing a cautionary claim: asymptotically "unscalable" algorithms might very well outperform "scalable" ones over architecture sizes of practical interest, unless care is taken in the analysis to account for second-order effects, as well as the interplay of the all-too-often neglected constants.

The rest of this paper is organized as follows. In Section 2 we give a precise definition of the two 2-D DWT algorithms (Standard and Non-standard forms), as well as of the partitioning schemes and of our notational conventions. In Section 3 we elaborate on the parallel machine model, including both the execution and communication times. Section 4 deals with the definition of the four *parallel* algorithms, and the derivation of their respective parallel execution times on the CT-routed Mesh. In Section 5 we define the isoefficiency metric, and derive the scalability properties of the algorithms on the CT-routed Mesh. In Section 6 we briefly give the scalability results on the CT-routed Hypercube network, and on the SF-routed Mesh. Finally, in Section 7 we summarize our findings.

2. 2-D DWT Algorithms and Partitioning Schemes

2.1. Standard and Non-standard 2-D DWT Algorithms

According to Beylkin et al. [9] the Non-standard (NS) form of the 2-D DWT results from the recursive application of a three-component operator; borrowing notation in [9], these three components are α^j, β^j and γ^j. Fig. 1 (a) shows the usual arrangement of the three decomposition components on a 3-octave tree. The output matrix of a NS-form decomposition is characterized by having three equal-size components at a single-dimensional octave level j; for instance, in Fig. 1 (a), the components at octave $j = 2$ are α^2, β^2 and γ^2.

The Standard form of the 2-D DWT, on the other hand, is a complete decomposition over the set of all possible pairs of separable wavelet basis functions (ψ_{j_1}, ψ_{j_2}). Fig. 1 (b) shows the usual arrangement of rectangular quadrants according to the octave *pairs* (j_1, j_2). In the S-form, a particular component of the output matrix is classified on a two-dimensional octave plane; for example, in Fig. 1 (b) quadrant (j_1, j_2) = (3, 1) represents the projection of the input on the subspace defined by basis functions $< \psi_3, \psi_1 >$. (Note that octave index j_1 is horizontal and j_2 vertical, opposite to element indices n_1, n_2 in Fig. 1.)

Computationally, there are several differences between the S- and NS-forms. Computation of an NS-form decomposition relies on the *pyramidal algorithm*, a now well-known iterative procedure described by Mallat in [4]. Fig. 1 (a) shows the iterative structure involved in applying the three-element α^j, β^j, γ^j operator first to the entire input matrix, and subsequently to the upper-left hand quadrant. This quadrant is the result of a low-pass filtering operation on both, the rows and the columns, of the input matrix at octave j. Fig. 1 (c) shows the block-diagram of the 2-D DWT algorithm in NS-form.

Formally, if $H(n)$ and $G(n)$ are 1-dimensional high- and low-pass wavelet filters respectively; j the octave index; and $S_{xy}^j(n_1, n_2)$ denotes the wavelet component at octave j obtained from a 1-D row (horizontal) convolution and decimation by two with filter x, and a 1-D column (vertical) convolution and decimation by two with filter y; then, at octave j the four wavelets components are given by

$$S_{GG}^j(n_1, n_2) = \sum_{k_1} \sum_{k_2} S_{GG}^{j-1}(k_1, k_2) G(2n_2 - k_2) G(2n_1 - k_1), \tag{1}$$

$$S_{GH}^j(n_1, n_2) = \sum_{k_1} \sum_{k_2} S_{GG}^{j-1}(k_1, k_2) G(2n_2 - k_2) H(2n_1 - k_1), \tag{2}$$

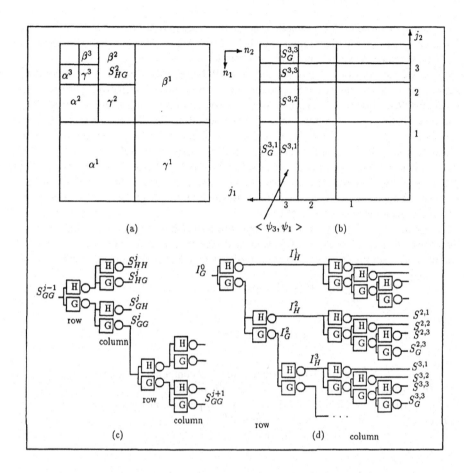

Figure 1. Non-standard and Standard 2-D DWT algorithm. (a) and (b) show the arrangement of the wavelet decomposition components for the NS- and S-forms, respectively. (c) shows one recursion of the NS algorithm in block diagram, and (d) shows a 3-octave example of the S-form (excluding the last branch). H, G are the high- and low-pass wavelet filters, and circles denote decimation by two.

$$S^j_{HG}(n_1, n_2) = \sum_{k_1} \sum_{k_2} S^{j-1}_{GG}(k_1, k_2) H(2n_2 - k_2) G(2n_1 - k_1), \tag{3}$$

$$S^j_{HH}(n_1, n_2) = \sum_{k_1} \sum_{k_2} S^{j-1}_{GG}(k_1, k_2) H(2n_2 - k_2) H(2n_1 - k_1). \tag{4}$$

For example, in Fig. 1 (a) component β_2 corresponds to $S^2_{HG}(n_1, n_2)$, and similarly α^j, γ^j correspond to $S^j_{GH}(n_1, n_2)$ and $S^j_{HH}(n_1, n_2)$ respectively. Clearly, as these are separable wavelet spaces, one can isolate row from column convolutions and decimations as well as interchange their order. All summations are in the range $\max(0, 2n_i - L + 1) \leq k_i \leq 2n_i$, $i = 1, 2$, where L is the filter length of both H and G. We denote the input to the algorithm as $M \times M$ matrix $\mathbf{A}$ with elements $a_{i,j}$, so that $S_{GG}^0(n_1, n_2) = a_{n_1, n_2}$.

There are two ways of computing the 2-D DWT algorithm S-form. One is by direct application of a row and column operator to input matrix $\mathbf{A}$. Alternatively, one may first obtain the NS-form, and then apply a transform on all the columns of its β^j components, and on all rows of its α^j components. (From which it is clear that the γ^j components of the NS-form appear identically along the main diagonal of the S-form.) We adopt the first direct approach, having the advantage of allowing us to separate the algorithm into two independent phases: (a) a one-dimensional row (or column) transformation on the input matrix, resulting in a full-octave decomposition along one octave dimension j_1 (j_2); and (b) a one-dimensional column (or row) transformation identical to step 1, resulting in a full-octave decomposition along j_2 (j_1).

Let $I_x^{j_1}(n_1, n_2)$ be the intermediate component resulting from a row (horizontal) convolution with filter x ($x = H, G$), followed by decimation by two, at octave j_1. Then, the components resulting from a *full*, $j_1 = J$ row decomposition summing on k_2 are given by:

$$I_G^{j_1}(n_1, n_2) = \sum_{k_2} I_G^{j_1-1}(n_1, k_2) G(2n_2 - k_2), \tag{5}$$

$$I_H^{j_1}(n_1, n_2) = \sum_{k_2} I_G^{j_1-1}(n_1, k_2) H(2n_2 - k_2), \tag{6}$$

for $1 \leq j_1 \leq J$. Now let $S^{j_1, j_2}(n_1, n_2)$ denote the final component at octave pair (j_1, j_2), and $S_G^{j_1, j_2}(n_1, n_2)$ denote the components which are the output from the last low-pass filters G.[2] As Fig. 1 (d) indicates, $S^{j_1, j_2}(n_1, n_2)$ is obtained by performing a j_2-octave decomposition (summing on k_1) for each intermediate term $I_x^{j_1}(n_1, n_2)$, $1 \leq j_1 \leq J$ and $x = H, G$. For instance, component $S^{3,1}(n_1, n_2)$ shown in Fig. 1 (d) is obtained via intermediate component $I_H^3(n_1, n_2)$.

2.2. Data Partitioning

One of the central issues in parallel computing is data partitioning, dealing with the assignment and distribution of data to processors. The partitioning strategy (at the beginning of the algorithm as well as during its operation) has enormous impact on parallel system performance, mainly because data partitioning dictates communication requirements, which in turn account for the bulk of inefficiencies. We consider two commonly used partitioning schemes, the *Checkerboard*- (CP) and the *Stripped*- (SP) partitions [23].

Consider the zero-indexed $M \times M$ matrix $\mathbf{A}$, with elements denoted as $a_{i,k}$, and a set of P processors, where the $(x, y)^{th}$ processor is denoted as $PE_{x,y}$ for $0 \leq x, y \leq \sqrt{P} - 1$. Fig. 2 (a) shows the assignment of matrix elements $a_{i,k}$ to processors in CP. We denote by $\mathbf{A}_{x,y}$ the square sub-matrix consisting of elements $a_{i,k}$ such that $\frac{Mx}{\sqrt{P}} \leq i \leq \frac{Mx}{\sqrt{P}} + \frac{M}{\sqrt{P}} - 1$, and $\frac{My}{\sqrt{P}} \leq k \leq \frac{My}{\sqrt{P}} + \frac{M}{\sqrt{P}} - 1$. For example, in the right part of Fig. 2 (a) we show $\mathbf{A}_{0,0}$. Without loss of generality, we assume that $\sqrt{P}$ and $M/\sqrt{P}$ are integers.

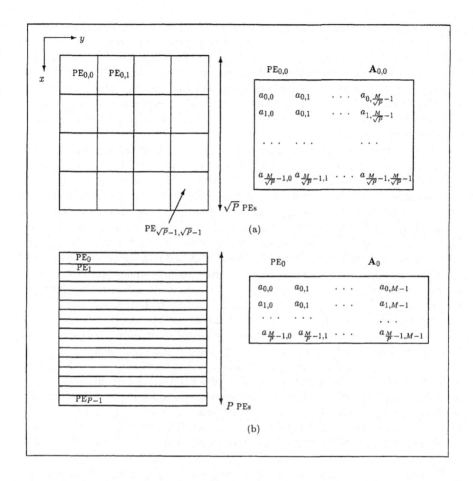

Figure 2. Partitioning Schemes. (a) checkerboard partitioned matrix among a set of P processors, and (b) row stripped partitioned matrix.

Similarly, we denote as A_x the sub-matrix consisting of elements $a_{i,k}$ such that $\frac{Mx}{P} \leq i \leq \frac{Mx}{P} + \frac{M}{P} - 1$ and $0 \leq k \leq M - 1$, defining the mapping in SP as shown in Fig. 2 (b).

We define as *Processor Granularity* the ratio of the total number of elements in the input matrix to the number of processors P. Note that for both, CP and SP schemes, processor granularity is equal to M^2/P.

3. Performance Analysis Model

The following assumptions hold for the performance model used throughout this paper. Total parallel execution time T_p consists of a computation component T_{comp}, and a communication component T_{comm}. T_{comp} is the total time needed for every computation (addition and multiplication) involved in the algorithm; and for normalization purposes, we let t_c be the time of one multiply and add operation. The data type of the operands (floating point, integer) is immaterial to the model, as long as t_c is constant. In this computation time model, we are ignoring memory-hierarchy effects. A linear relationship between the parallel execution time and the number of computations has been experimentally observed over a wide-range of architectures, and provides a convenient model as long as its limitations are understood.

Overall communication time T_{comm} depends heavily on the underlying interconnection network's topology, and the mechanisms governing message interchanges. Processor networks are often modeled as a graph $G = (V, E)$ with vertices V corresponding to processors, and edges E to bi-directional communication links. We adopt the *single-port* model, where processors may only send or receive a message over one communication link at a time. Single-port models have proven to be effective in dealing with a large class of machines, providing a pessimistic model on communication performance. The MasPar MP-1 and MP-2, Thinking Machines CM-2 and CM-5, and Intel's Paragon, are examples of single-port architectures. In *k-port* models, a processor may send or receive k messages at unison over k links. Parsytec's PowerPlus, for example, based on Inmos' Transputer with 4 communication links fits into the 4-port model. The 2-D DWT algorithms seem not able to effectively utilize more than a single communication port at a time.

We consider two message exchange mechanisms, *virtual-cut-through* (CT) and *store-and-forward* (SF) routing. In SF-routed networks, processors with no direct link interchanging a size-m message over l-links experience a communication delay of $T_{comm}^{SF} = t_s + l(mt_w + t_h)$, where t_s is the time incurred in establishing a connection (setting-up data structures, opening channels); t_w is the reciprocal of the channel bandwidth in seconds per unit of information (a size-m message has m units, i.e. m bytes for 8-bit pixel data); and t_h is the time delay incurred in transmission over a single link, or hop time. All intermediate nodes participating in a SF-routed connection store the entire message, consequently requiring at least size-m buffers.

CT-routed networks rely on a form of pipelining for reducing transmission delays over multiple-length connections. Rather than storing an entire size-m message, in CT-routing the source processor segments messages into f smaller units called *flits* of size m/f. Therefore, intermediate processors need only maintain buffers of size m/f, reducing communication time from $\Theta(lm)$ to $\Theta(l+m)$. The communication delay of CT-routed networks is given by $T_{comm}^{CT} = t_s + mt_w + lt_h$.

SF-routed networks are gradually being replaced by CT-routing due to the shorter delays and smaller buffers inherent in CT. For completeness, we present our results for both CT- and SF-routed networks.

3.1. Mesh and Hypercube Processor Networks

In a $X \times Y$ 2-D Mesh-connected network, processors are labeled with indices x, y with $0 \leq x < X, 0 \leq y < Y$, and there exists a link between PE_{x_1, y_1} and PE_{x_2, y_2} if and only if $(|x_1 - x_2| \leq 1$ and $|y_1 - y_2| \leq 1)$ and $(|x_1 - x_2| = 0$ or $|y_1 - y_2| = 0)$. A Mesh with toroidal wrap-around connections has the further property that $PE_{X-1,y}$ is connected to $PE_{0,y}$, and so is $PE_{x,Y-1}$ to $PE_{x,0}$. All processors in a toroidal 2-D Mesh network have exactly four links.

A d-dimensional Hypercube network may be described by a binary labeling of processors with indices $B = (b_{d-1}, b_{d-2}, \ldots, b_1, b_0)$, where b's stand for bits with b_{d-1} being the most significant bit. In a d-dimensional Hypercube, processor PE_{B_1} has a connection with PE_{B_2} if and only if the Hamming distance between B_1 and B_2 is equal to one, i.e. they differ in exactly one bit position. The number of links in every processor of a d-dimensional Hypercube is d.

4.   Parallel Algorithms

In this section we derive the parallel execution time T_p^{NS-CP} of the NS-CP algorithm, briefly describe the steps involved in obtaining T_p^{S-SP} for S-SP, and give the parallel times of S-CP and NS-SP.

4.1.   Problem Size and Serial Execution Times

Scalability analysis is concerned with determining how well an algorithm makes use of increasing resources in a parallel machine. In order to establish a meaningful comparison for an algorithm running on *machines* with different sizes, one must consider and quantify *problem* size. The problem size is the sum of all the computations in algorithm C, and may be expressed either as a number C or as the serial execution time $T_s = Ct_c$.

The total number of computations in both 2-D DWT forms, S and NS, is a number that depends on the $M \times M$ input matrix **A** as well as on decomposition depth J. Problem size C is therefore a function of two variables, M and J. Scaling the DWT holding J constant results in a trivial problem, and in addition would not fairly reflect the increase in decomposition depth J that becomes available after M has increased. Suppose, for instance, that a problem with $M \times M$ input matrix and J levels is to be scaled up to $2M \times 2M$. It is clearly necessary to balance the larger input matrix with a deeper octave decomposition $J + 1$, since the larger matrix allows one extra octave. Let γ be a constant such that $0 < \gamma < 1$. Also, let the highest octave in a problem be given by $J = \lceil \gamma \log_2 M \rceil$.[3] Now, the problem size of both DWT forms becomes a function of M and γ. In scaling up a problem from M_1 to M_2, however, we will hold γ *constant*, thus effectively making the problem size a function only of M.

The total number of multiply-add operations needed to decompose an $M \times M$ matrix into J octaves in the NS-form is calculated as follows. We assume that $M >> L$, where L is the filter length. For the first octave, M rows of size M are convolved with filters H, G and decimated by two. Rather than decimating *after* filtering—resulting in redundant work—decimation is performed before filtering. In computation count, two filters preceded by decimation are equivalent to a single filter with no decimation, namely ML operations per row. As a result, the total number of computations for all rows is M^2L. (We disregard the $L - 1$ output points at the tail of convolution.) Total operation count, then, for both rows and columns is $2M^2L$.

Computation of octave two proceeds identically as octave one, except the input matrix is now $M/2 \times M/2$, resulting in a total of $\frac{1}{2}M^2L$ operations. Proceeding likewise and adding a total of J octaves, one gets

$$C^{NS} = \sum_{j=1}^{J} \frac{2M^2L}{4^{j-1}} = \frac{8}{3}M^2L\left(1 - \frac{1}{4^J}\right). \tag{7}$$

The serial execution time is modeled as $T_s^{NS} = C^{NS}t_c$, and this quantity is taken as the baseline for computing the efficiency of the NS-form algorithms.

Following similar arguments it can be shown that the S-form algorithm has a total computation count (work) of

$$C^S = 4M^2L\left(1 - \frac{1}{2^J}\right). \tag{8}$$

4.2.   NS-CP on a $\sqrt{P} \times \sqrt{P}$ Mesh

Let input matrix $\mathbf{A}$ be distributed among P processors in CP as shown in Fig. 2 (a). Let the set of P processors be connected in a Mesh network as described in Section 3.1 and labeled as shown in Fig. 3 (a).[4]

Schematically, the operation of NS-CP is as follows. Every processor performs two 1-D local convolutions with decimation by two (both filters H and G) on every one of the $M/\sqrt{P}$ rows of length $M/\sqrt{P}$. Processor $PE_{0,0}$, for example, holding the matrix block shown in Fig. 2 (a) performs convolution and decimation on rows with elements $\{a_{i,0}, a_{i,1}, \ldots, a_{i,\frac{M}{\sqrt{P}}-1}\}, 0 \le i < \frac{M}{\sqrt{P}}$. Clearly, all processors not lying along contour $y = 0$ will require certain inputs from their neighbors at column $y - 1$.

After all first-octave convolutions with decimation are complete, an identical operation is performed on the input matrix *columns*, resulting in the same number of operations and messages. Octave two proceeds likewise, but with input $S_{GG}^1(n_1, n_2)$.

The NS-CP parallel algorithm is shown in Table 1, and it is a *Single Program Multiple Data* (SPMD) algorithm in which all processors execute identical code on local memory data. Processor $PE_{x,y}$ initially holds block $\mathbf{A}_{x,y}$ of input matrix $\mathbf{A}$ in variable $S_{GG}^0(n_1, n_2)$, as indicated in line 3. (Note that the zero-based indices n_1, n_2 are *local* to processor $PE_{x,y}$.) Consider, for example, the case $P = 16$ and $M = 8$. Then $PE_{1,1}$ initially holds $a_{2,2}, a_{2,3}, a_{3,2}, a_{3,3}$ in $S_{GG}^0(0, 0), S_{GG}^0(0, 1), S_{GG}^0(1, 0), S_{GG}^0(1, 1)$. Let the variable M' be the square root of processor granularity at octave j, and S_A be a binary variable determining the state of $PE_{x,y}$. If $S_A = 1$ then $PE_{x,y}$ is active, if $S_A = 0$ then $PE_{x,y}$ is inactive. Parameter j_0 will be defined below; but meanwhile, let us assume that M and P are such that $J \le j_0$, so that lines 10 to 12 in the NS-CP algorithm of Table 1 are not executed.

Total number of multiply-add computations is determined with the aid of the *Dependence Graph* (DG) of 1-D convolution with decimation by two [27]. The DG is a representation of an algorithm as a graph where vertices represent computations (multiply-add), and

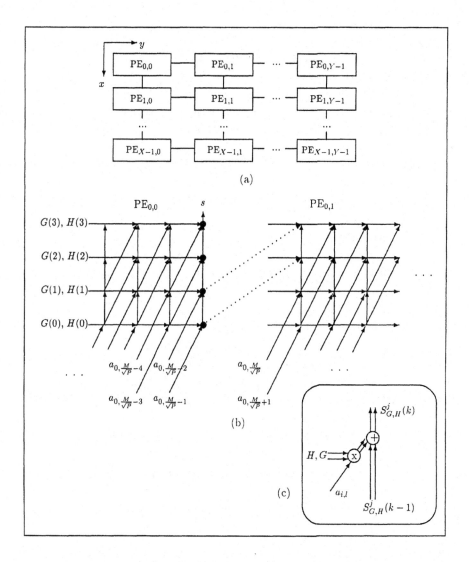

Figure 3. (a) Mesh processor network; (b) Dependence Graph (DG) of 1-D convolution and decimation at the boundary of $PE_{0,0}$ and $PE_{0,1}$, with $L = 4$; (c) multiply-add operations in DG node.

edges data communication. Fig. 3 (b) shows a section of the DG of 1-D convolution with decimation for single row $i = 0$ at the boundary of $PE_{0,0}$ and $PE_{0,1}$, with $L = 4$ tap filters H and G; Fig. 3 (c) shows the two multiply-add computations taking place in every DG vertex. Vertical edges pointing north represent the accumulation of partial results; horizontal edges pointing east represent filter coefficients G, H which are local to each PE; and

Table 1. Pseudo-code of NS-CP parallel algorithm.

1. Let $M' \equiv M/\sqrt{P}2^{j-1}$, $j_0 \equiv \log(M/\sqrt{P})$, and $\rho(x) \equiv x \bmod 2^{j-j_0}$

2. n_1, n_2 are zero-based local indices at $PE_{x,y}$

3. Initially $(j = 1)$, $PE_{x,y}$ holds $\mathbf{A}_{x,y}$ in local $S^0_{GG}(n_1, n_2)$

4. Let $s^j_G(n_1, n_2) = 0$, $s^j_H(n_1, n_2) = 0$ for $1 \leq j \leq J$ be temporary variables, and

5. $\qquad S^j_{GG} = S^j_{GH} = S^j_{HG} = S^j_{HH} = 0 \; \forall n_1, n_2$ and $j \neq 0$

6. set $S_A = 1$ in all PEs

7. for $j = 1$ to J

8. $\qquad$ PEs in S_A are active

9. $\qquad$ if $j > j_0$

10. $\qquad\qquad$ $M' = 2$

11. $\qquad\qquad$ Consolidate S^j_{GG}

12. $\qquad\qquad$ if $\rho(x) = 2^{j-j_0} - 1$ and $\rho(y) = 2^{j-j_0} - 1$ then $S_A = 1$

13. $\qquad$ **MR**: $PE_{x,y+1}$ sends to $PE_{x,y+1}$ $\{S^{j-1}_{GG} \mid 0 \leq n_1 \leq M' - 1, M' - L + 2 \leq n_2 \leq M' - 1\}$

14. $\qquad$ **R**: for $0 \leq n_1 \leq M' - 1$

15. $\qquad\qquad$ for $0 \leq n_2 \leq M'/2 - 1$

16. $\qquad\qquad\qquad$ for $\max(0, 2n_2 - L + 1) \leq k_2 \leq 2n_2$

17. $\qquad\qquad\qquad\qquad$ $s^j_G(n_1, n_2) \leftarrow s^j_G(n_1, n_2) + S^{j-1}_{GG}(n_1, k_2)G(2n_2 - k_2)$

18. $\qquad\qquad\qquad\qquad$ $s^j_H(n_1, n_2) \leftarrow s^j_H(n_1, n_2) + S^{j-1}_{GG}(n_1, k_2)H(2n_2 - k_2)$

19. $\qquad$ **MC**: $PE_{x,y}$ sends to $PE_{x+1,y}$ $\{s^j_{G,H} \mid M' - L + 2 \leq n_1 \leq M' - 1, 0 \leq n_2 \leq M' - 1\}$

20. $\qquad$ **C**: for $0 \leq n_2 \leq M' - 1$

21. $\qquad\qquad$ for $0 \leq n_1 \leq M'/2 - 1$

22. $\qquad\qquad\qquad$ for $\max(0, 2n_1 - L + 1) \leq k_1 \leq 2n_1$

23. $\qquad\qquad\qquad\qquad$ $S^j_{GG}(n_1, n_2) \leftarrow S^j_{GG}(n_1, n_2) + s^j_G(k_1, n_2)G(2n_1 - k_1)$

24. $\qquad\qquad\qquad\qquad$ $S^j_{GH}(n_1, n_2) \leftarrow S^j_{GH}(n_1, n_2) + s^j_G(k_1, n_2)H(2n_1 - k_1)$

25. $\qquad\qquad\qquad\qquad$ $S^j_{HG}(n_1, n_2) \leftarrow S^j_{HG}(n_1, n_2) + s^j_H(k_1, n_2)G(2n_1 - k_1)$

26. $\qquad\qquad\qquad\qquad$ $S^j_{HH}(n_1, n_2) \leftarrow S^j_{HH}(n_1, n_2) + s^j_H(k_1, n_2)H(2n_1 - k_1)$

slanted edges pointing northeast represent the propagation of inputs $a_{i,k}$. The dotted arrows represent those input points which must be transferred from $PE_{0,0}$ to $PE_{0,1}$.

For illustration, consider the output labeled s in Fig. 3 (b), produced at the upper-right DG node on $PE_{0,0}$. In 1-D convolution by filter G (or H) followed by decimation, output s is computed by the summation of four partial products

$$a_{0,\frac{M}{\sqrt{P}}-4}G(3), \quad a_{0,\frac{M}{\sqrt{P}}-3}G(2), \quad a_{0,\frac{M}{\sqrt{P}}-2}G(1), \quad a_{0,\frac{M}{\sqrt{P}}-1}G(0),$$

which are shown as bold vertices. Note that the accumulation of these partial results is performed locally at $PE_{0,0}$, along the edges pointing north and ending in result s. Also note that the two data tokens requiring transfer from $PE_{0,0}$ to $PE_{0,1}$ are *input* data points $a_{0,\frac{M}{\sqrt{P}}-2}$, and $a_{0,\frac{M}{\sqrt{P}}-1}$, for the case $L = 4$. For the general case, $PE_{x,y}$, $y \neq 0$, will receive a message of size $\frac{M}{\sqrt{P}}(L - 2)$, since there are $\frac{M}{\sqrt{P}}$ rows, and each row needs to communicate $L - 2$ input elements.[5] In Fig. 3 (b), Processor $PE_{0,1}$ will receive from $PE_{0,0}$ elements $\{a_{i,\frac{M}{\sqrt{P}}+2-L}, \ldots, a_{i,\frac{M}{\sqrt{P}}-1}\}$, for $0 \leq i < \frac{M}{\sqrt{P}}$. Finally, note that this DG clearly shows that the number of outputs resulting from convolution by *one* filter is half the number of input data points, due to decimation by two. As a result, the number of output points after convolution by *two* filters with decimation is equal to the number of input points.

4.2.1. T_p^{NS-CP} with Processor Granularity Grater than 4

The number of operations in octave one, then, is as follows. There are $ML/\sqrt{P}$ computations per processor per row, including both filters and decimation. Therefore, the total for $M/\sqrt{P}$ rows and columns is $2M^2L/P$. In the NS-CP algorithm shown in Table 1, lines 14 to 18 and 20 to 26 labeled **R** (row), and **C** (column) detail the specifics of the row and column convolutions and decimations respectively. As $M >> L$, we do not consider the $L(L - 2)M/\sqrt{P}$ edge computations at $PE_{x,Y-1}$, $0 \leq x < X$. The ratio of edge computations to C is $(L - 2)\sqrt{P}/2M$, which is less than $1/1000$ for typical numbers.

The communication load consists of two, size-$(L - 2)M/\sqrt{P}$ messages per processor (rows and columns). All messages occupy a link along the y-direction from $PE_{x,y}$ to $PE_{x,y+1}$, so there is no link contention. Lines 13 and 19 of the NS-CP algorithm in Table 1 labeled **MR** (message row), and **MC** (message column) show the specific elements that constitute these messages.

At the conclusion of octave one, the four $M/2 \times M/2$ sub-matrices $S_{GG}^1(n_1, n_2)$, $S_{GH}^1(n_1, n_2)$, $S_{HG}^1(n_1, n_2)$ $S_{HH}^1(n_1, n_2)$ are again checkerboard-partitioned as the input matrix **A** was. Therefore, octave two proceeds identically as octave one, but with an $M/2 \times M/2$ matrix $S_{GG}^1(n_1, n_2)$, resulting in $M^2L/2P$ computations and two, size-$(L - 2)M/2\sqrt{P}$ messages per processor.

As the size of $S_{GG}^j(n_1, n_2)$ decreases for increasing j, there is an octave at which there will only be 4 input points per processor, or $M' = 2$. Let j_0 be the octave where processor granularity equals 4, which is given by $j_0 = \log(M/\sqrt{P})$.

Summing over j for $J \leq j_0$, one obtains

$$C^{NS-CP}(J \leq j_0) = \frac{8LM^2}{3P}\left(1 - \frac{1}{4^J}\right), \tag{9}$$

which is precisely equal to the serial operation count C^{NS} divided by P, as given in Eqn. (7). Therefore, parallel time for $J \leq j_0$ is given by $T_{comp}^{NS-CP} = \frac{8}{3}Lt_c(1 - \frac{1}{4^J})\frac{M^2}{P}$.

Communication requirements for $J \leq j_0$ are equal for CT- and SF-routed networks, since all communication is distance-1. Adding the contribution of $2J$ messages of size $(L - 2)M/\sqrt{P}2^{j-1}$ over all j for $J \leq j_0$ gives a total communication time of

$$T_{comm}^{NS-CP}(J \leq j_0) = 2J(t_s + t_h) + 4(L - 2)\left(1 - \frac{1}{2^J}\right)t_w\frac{M}{\sqrt{P}}. \tag{10}$$

4.2.2. T_p^{NS-CP} with Processor Granularity of 4

Let us now relax the assumption that $J \le j_0 = \log(M/\sqrt{P})$. The equations for execution and communication times derived above are valid as long as all communication is distance-1, occuring when $1 \le j \le j_0$. However, when $j > j_0$, there are three major changes in the algorithm's operation. First, the set of active PEs is no longer the entire mesh; in line 12 of Table 1, variable S_A is updated to the current active processor set at octave j, based on condition $\rho(x) = 2^{j_0-1}$, with $\rho(x)$ defined as $\rho(x) \equiv x \bmod 2^{j-j_0}$ (and similarly for $\rho(y)$). Second, all communication takes place among processors that are further away than distance-1. In lines 13 and 19, the transmission distance is given by 2^{j-j_0}, instead of one. And third, there is an extra step (line 11) for consolidating the outputs from octave j generated at four PE's into a single PE. This last step is required because at the end of every octave j beyond j_0, there is only a single element S_{GG} per active PE, and thus it becomes necessary to merge three elements before starting octave $j + 1$. We omit the straight-forward details involved in the consolidation step due to space limitations. For details, refer to [21].

The number of computations for both rows and columns at octave j for $j > j_0$ is constant, and leads to $T_{comp}^{NS-CP}(j) = 4Lt_c$. The row and column communication requirements per octave j consist of two, size-$(L-2)$ messages sent across distance-2^{j-j_0} links. To account for data consolidation at the octave interface, we adjust message length by adding one additional element for a message size of $L - 1$. As a result, the communication time for the CT-routed Mesh at octave j in the interval $j > j_0$ is given by

$$T_{comm}^{NS-CP}(j) = 2t_s + 2(L-1)t_w + 2^{j-j_0+1}t_h, \tag{11}$$

and the total communication time over all octaves in $j_0 < j \le J$ by

$$\sum_{j=j_0+1}^{J} T_{comm}^{NS-CP}(j) = 2(J - j_0)(t_s + (L-1)t_w) + 2(2^{J-j_0+1} - 2)t_h.$$

Finally, adding the contribution of $T_{comp}^{NS-CP}(j \le j_0)$, $T_{comm}^{NS-CP}(j \le j_0)$, $T_{comp}^{NS-CP}(j > j_0)$, and $T_{comm}^{NS-CP}(j > j_0)$, NS-CP's total parallel execution time in a $\sqrt{P} \times \sqrt{P}$ CT-routed Mesh network is given by

$$T_p^{NS-CP} = \begin{cases} \frac{8}{3}Lt_c(1 - \frac{1}{4^J})\frac{M^2}{P} + 2J(t_s + t_h) + 4(L-2)(1 - \frac{1}{2^J})t_w\frac{M}{\sqrt{P}} & \text{for } J \le j_0 \\[2mm] \frac{8}{3}Lt_c(1 - \frac{1}{4^{j_0}})\frac{M^2}{P} + 2j_0(t_s + t_h) \\ +4(L-2)(1 - \frac{1}{2^{j_0}})t_w\frac{M}{\sqrt{P}} + 4Lt_c(J - j_0) \\ +2(J - j_0)(t_s + (L-1)t_w) + 2(2^{J-j_0+1} - 2)t_h & \text{for } J > j_0 \end{cases} \tag{12}$$

Table 5 lists the terms in the expression for T_p^{NS-CP} along with the associated constants. Although most terms in Table 5 and Eqn. (12) admit slight further simplification, they have been left as shown for clarity. Since J and j_0 are both functions of M and P, all listed terms can be expressed in terms of M and P only, where for example, term (a) simplifies to $\frac{M^2}{P} - 1$.

4.3.   S-SP, S-CP and NS-SP

The standard form of the DWT algorithm with stripped partitioning is characterized by the property that computations may be executed locally with zero inter-processor communications in a row phase called $\mathbf{F}_r$, and a column phase $\mathbf{F}_c$. All communications are performed in a single phase $\mathbf{F}_{comm}$, implemented as an all-to-all broadcast. S-SP has an important advantage over the other three DWT algorithms from the point of view of programming and portability. Writing code for the purely computational phases is no different than programming a single-processor machine, and the all-to-all personalized communication step may be implemented using pre-existing library primitives tuned for native machines.

Let input matrix $\mathbf{A} = (a_{i,k})$ be initially distributed among P processors in P sub-matrix blocks $\mathbf{A}_x$, as defined in Section 2.2. PE's are indexed PE_x, $0 \le x \le P - 1$, as shown in Fig. 2 (b). The data structure local to PE_x holding $\mathbf{A}_x$ is denoted $I_G^0(n_1, n_2)$. (Note that, as in NS-CP, zero-based indices n_1, n_2 are local to PE_x.) The S-SP algorithm is summarized in Table 2.

Each one of the $\frac{M}{P}$ rows in $I_G^0(n_1, n_2)$ is an input to the full, 1-D J-octave decomposition as described by Eqns. (5) and (6). Phase $\mathbf{F}_r$ at PE_x produces blocks

$$\{I_G^{j_1=J}(n_1, n_2), I_H^J(n_1, n_2), \cdots, I_H^2(n_1, n_2), I_H^1(n_1, n_2)\} \text{ for } 1 \le j_1 \le J,$$

as shown in lines 3 through 8 of Table 2. For clarity, we now make the assumption that P is a power of two.[6] Let the output blocks $I_H^{j_1}$ and I_G^J, $1 \le j_1 \le J$, be arranged in a single data structure $I'(n_1, n_2)$ such that $I_H^{j_1}$ will appear in $I'(n_1, n_2)$ for $\frac{M}{2^{j_1}} \le n_2 < \frac{M}{2^{j_1-1}}$, and I_G^J will appear in $I'(n_1, n_2)$ for $0 \le n_2 < \frac{M}{2^J}$, both with $0 \le n_1 < \frac{M}{P} - 1$, as shown in line 9 of Table 2. Thus,

$$I'(n_1, n_2) = \left[\; I_G^J \; \middle| \; I_H^J \; \middle| \cdots \middle| \; I_H^2 \; \middle| \; I_H^1 \; \right]. \tag{13}$$

Fig. 4 (a) shows an example of a three-octave arrangement in $I'(n_1, n_2)$.

Next, in phase $\mathbf{F}_{comm}$, PE_x composes $P - 1$ messages consisting of blocks in $I'(n_1, n_2)$. PE_x sends a message to $PE_{x'}$ with the elements in $I'(n_1, n_2)$ such that $0 \le n_1 \le \frac{M}{P} - 1$, and $\frac{x'M}{P} \le n_2 \le \frac{x'M}{P} + \frac{M}{P} - 1$ (line 11). Note that the block for which $x' = x$ in $I'(n_1, n_2)$ stays at PE_x. The four messages in a 4-PE example are indicated in Fig. 4 (a).

At the end of phase $\mathbf{F}_{comm}$, PE_x has received $P - 1$ messages, one from every other PE. These messages are locally arranged in data structure $\tilde{I}(n_1, n_2)$, so that the message received from $PE_{x'}$ is written in $\frac{x'M}{P} \le n_1 \le \frac{x'M}{P} + \frac{M}{P} - 1$, and $0 \le n_2 \le \frac{M}{P} - 1$, as shown in Fig. 4 (b), and in line 12 of Table 2. $\tilde{I}(n_1, n_2)$ now holds, along n_1, the columns of the intermediate results.[7] Phase $\mathbf{F}_c$ proceeds as phase $\mathbf{F}_r$, but now the summations in Eqns. (5) and (6) are performed over k_1.

At the completion of phase $\mathbf{F}_c$, re-arrangement of $\tilde{I}$ as I' in Eqn. 13 will result in the final location of the decomposition results S^{j_1,j_2}, $1 \le j_1, j_2 \le J$, as shown in Fig. 4 (c) for the 4-PE example. Notice there is a direct correspondence between the location of the results in Fig. 4 (c) and the decomposition shown in Fig. 1 (b): The components shown in Fig. 4 (c) for PE_1 are exactly those corresponding to the second branch (I_H^2) in the block diagram of Fig. 1 (d); the results of the first branch (I_H^1) are split among PE_2 and PE_3; and the last two branches (I_H^3 and I_G^3) are in PE_0.

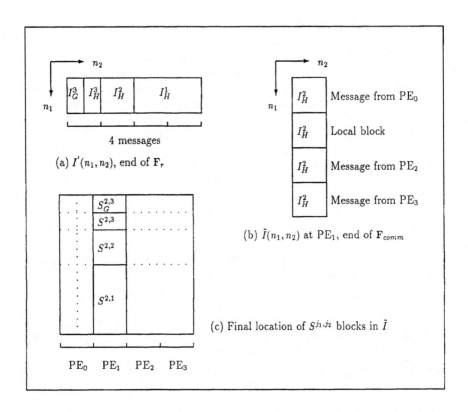

Figure 4. (a) Arrangement of $I_H^{j_1}$ and I_G^{J} blocks in $I'(n_1, n_2)$ after phase $\mathbf{F}_r$ of S-SP, with $J = 3$. (b) Messages received at PE$_1$ after phase $\mathbf{F}_{comm}$, with $P = 4$. (c) Final location of results at PE$_1$.

A similar computation count as for NS-CP shows that S-SP requires a total of $\frac{4LM^2}{P}(1 - \frac{1}{2^J})$ multiply-add operations, which is exactly equal to $\frac{C_s}{P}$ (see Eqn. (8)).

The running time for the all-to-all personalized communication operation is well-understood; due to lack of space we only give here the results (see for example [23]). On both, CT- and SF-routed Meshes, all-to-all personalized communication on size m messages requires time of $(2t_s + t_w m P)(\sqrt{P} - 1)$, while in CT- and SF-routed Hypercubes it requires $(t_s + mt_w)(P - 1) + \frac{1}{2}t_h P \log P$, and $(t_s + \frac{1}{2}t_w m P) \log P$ respectively. The total parallel running time of S-SP is listed in tabular form in Table 5, where $m = M^2/P^2$.

The parallel running times for S-CP and NS-SP are also given in Table 5. These two algorithms are not discussed in detail, but we give a schematic version for them written in pseudo-code in Tables 3 and 4.

The S-CP algorithm is very similar to NS-CP, especially in communication mechanics and message construction. Qualitatively, the main difference between them is a slight increase in message size in S-CP as compared to the message size in NS-CP.

Table 2. Pseudo-code of S-SP parallel algorithm.

1. n_1, n_2 are zero-based local indices at PE_x.

2. Initially, PE_x holds $\mathbf{A}_x$ in local $I_G^0(n_1, n_2)$

3. $\mathbf{F}_r$: for $j_1 = 1$ to J

4.   for $0 \leq n_1 \leq \frac{M}{P} - 1$

5.   for $0 \leq n_2 \leq M/2^{j_1} - 1$

6.   for $\max(0, 2n_2 - L + 1) \leq k_2 \leq 2n_2$

7. $I_G^{j_1}(n_1, n_2) \leftarrow I_G^{j_1}(n_1, n_2) + I_G^{j_1-1}(n_1, k_2)G(2n_2 - k_2)$

8. $I_H^{j_1}(n_1, n_2) \leftarrow I_H^{j_1}(n_1, n_2) + I_G^{j_1-1}(n_1, k_2)H(2n_2 - k_2)$

9. Arrange: $I'(n_1, n_2) \leftarrow \left[\, I_G^J \mid I_H^J \mid \cdots \mid I_H^2 \mid I_H^1 \,\right]$.

10. $\mathbf{F}_{comm}$: PE_x sends message to $PE_{x'}$, $x' \in [0, P-1]$, $x' \neq x$, with elements

11. $\{I'(n_1, n_2) \mid 0 \leq n_1 < \frac{M}{P}, \frac{x'M}{P} \leq n_2 < \frac{x'M}{P} + \frac{M}{P}\}$

12. PE_x arrange messages from $PE_{x'}$ in $\tilde{I}$ such that $\frac{x'M}{P} \leq n_1 < \frac{x'M}{P} + \frac{M}{P}$, $0 \leq n_2 < \frac{M}{P}$

13. $\mathbf{F}_c$: for $j_2 = 1$ to J

14.   for $0 \leq n_2 \leq \frac{M}{P} - 1$

15.   for $0 \leq n_1 \leq M/2^{j_2} - 1$

16.   for $\max(0, 2n_1 - L + 1) \leq k_1 \leq 2n_1$

17. $\tilde{I}^{j_1,j_2}(n_1, n_2) \leftarrow \tilde{I}^{j_1,j_2}(n_1, n_2) + \tilde{I}^{j_1,j_2-1}(k_1, n_2)G(2n_1 - k_1)$

18. $S_H^{j_1,j_2}(n_1, n_2) \leftarrow S_H^{j_1,j_2}(n_1, n_2) + \tilde{I}^{j_1,j_2-1}(k_1, n_2)H(2n_1 - k_1)$

Table 3. Brief Description of S-CP parallel algorithm.

1. Let $j_0 \equiv \log(M/\sqrt{P})$

2. Initially, $PE_{x,y}$ holds $\mathbf{A}_{x,y}$ of size $\frac{M}{\sqrt{P}} \times \frac{M}{\sqrt{P}}$

3. for $j_1 = 1$ to J

4.   row convolutions with decimation of $\frac{M}{\sqrt{P}}$ rows, with dist-1 and dist-2^{j-j_0} messages

5. for $j_2 = 1$ to J

6.   column conv. with decimation of $\frac{M}{\sqrt{P}}$ columns, with dist-1 and dist-2^{j-j_0} messages

Finally, the NS-SP algorithm is characterized by two phases. One having localized zero-length communications (row convolutions), and the other requiring distance-2^{j-j_0} communications (column convolutions) similar to those of NS-CP and S-CP. However, parameter j_0 in NS-SP is given by $j_0 = \frac{M}{P}$, which is *not* the same as that of NS-CP and S-CP. As will be shown in Section 5, NS-SP has considerably poorer scalability as compared to the other three algorithms for almost all architectures and input parameters. We include it here for comparison and completeness.

Table 4. Brief description of NS-SP parallel algorithm.

1. Let $j_0 \equiv \log(M/P)$
2. Initially, PE_x holds $\mathbf{A}_x$ of size $\frac{M}{P} \times M$
3. for $j = 1$ to J
4.   row convolutions with decimation of $\frac{M}{P}$ rows, no communication
5.   column convolutions with decimation of $\frac{M}{P}$ columns, dist-1 and dist-2^{j-j_0} messages

Table 5. Parallel running times for $J > j_0$ on a CT-routed Mesh. Left column lists all terms appearing in the execution time equations, and other columns list associated constants in each algorithm. Empty spaces in the columns mean that an algorithm does not contain that particular term.

	Term	NS-CP	S-SP	S-CP	NS-SP
(a)	$(1-\frac{1}{4^{j_0}})\frac{M^2}{P}$	$\frac{8}{3}Lt_c$			$\frac{8}{3}Lt_c$
(b)	$(1-\frac{1}{2^J})\frac{M^2}{P}$		$4Lt_c$		
(c)	$(1-\frac{1}{2^{j_0}})\frac{M^2}{P}$			$4Lt_c$	
(d)	$(1-\frac{1}{2^{j_0}})\frac{M}{\sqrt{P}}$	$4(L-2)t_w$			
(e)	$(1-\frac{1}{2^{j_0}})M$				$2(L-2)t_w$
(f)	$(\frac{1}{2^{j_0}}-\frac{1}{2^J})M$				$6Lt_c$ $2(L-1)t_w$
(g)	j_0	$2t_h - 4Lt_c$ $-2(L-1)t_w$		$2t_h$	t_h
(h)	$j_0\frac{M}{\sqrt{P}}$			$-4Lt_c-2t_w$	
(i)	J	$4Lt_c+2t_s$ $+2(L-1)t_w$		$2t_s$	
(j)	$J\frac{M}{\sqrt{P}}$			$4Lt_c$ $+2(L-1)t_w$	
(k)	2^{J-j_0+1}	$2t_h$		$2t_h$	t_h
(l)	$\sqrt{P}$		$2t_s$		
(m)	$\frac{M^2}{\sqrt{P}}$		t_w		
(o)	$\frac{M^2}{P}$		$-t_w$		
	Constant	$-4t_h$	$-2t_s$	$-4t_h$	$-2t_h$

5. Scalability of DWT Algorithms on a CT-routed Mesh

Parallel system performance is known to degrade with increasing computational resources, a fact frequently used for arguing against massively parallel processing. However, recently, researchers have found that under some conditions it is possible to maintain a fixed performance level with increasing resources as long as the input also grows in size. Scalability is a property of parallel systems reflecting the degree to which a problem size must increase in order to maintain constant performance.

A number of metrics for quantifying scalability have recently appeared in the literature. Among them we mention the slightly different versions of *scaled speedup* by Gustafson [28], Worley [29], and Sun and Ni [30]; the *equal performance condition* by Van-Catledge [31]; the *isospeed* measure by Sun and Rover [32]; and the *isoefficiency function* by Kumar et. al. [23]. Some of these metrics are complementary to one another, and their differences are mostly centered around *measuring* an algorithm's or a network's scalability. Most definitions agree on the basis for the approach to scalability, defining it as the ability of the combined algorithm and architecture to exploit increasing computational resources. Current debate is largely concerned with quantifying scalability, and it is not yet clear what constitutes a fair measure [33].

In this paper we use Kumar's isoefficiency metric, as it embraces several aspects of other metrics. With Kumar's isoefficiency it is not only possible to determine whether a system is linearly scalable,[8] but in addition, the specific function relating the input size to the number of processors necessary to maintain equal performance. For instance, in [32] it is proven that the isospeed metric reduces to the isoefficiency when the serial time execution model adopted is $T_s = C t_c$.

In our analysis we use a refined version of asymptotic isoefficiency. Although asymptotic performance broadly characterizes algorithmic behavior, it can be misleading if not carefully employed. It is very often the case that the constants determine an algorithm's performance *in specific regions* of problem and architecture sizes of interest, to a larger extent than asymptotics alone. That is, we use a *practical* form of scalability, taking into account the effect of the constants and their relations.

5.1. Definition of Isoefficiency

The *speedup* (S) of a parallel system on a P processor network is given by T_s/T_p, and the *efficiency* (E) by the ratio of speedup to P, or $E = T_s/PT_p$. It is clear that, as P grows, a system's efficiency, or its ability to provide increasing computational throughput, decays due to overhead consisting primarily of (a) inter-processor communication, and (b) idle processor time.

Let T_0 be the total amount of *overhead work* (time-processor product) incurred by the parallel system, expressed as $T_0 = PT_p - T_s$. Using this equation we get $E = \frac{1}{1+T_0/T_s}$. As a result, in order to maintain constant efficiency with increasing P, the overhead of the parallel system must not grow at a faster rate than the serial execution time. Since T_0 is a function of problem size and number of processors P, solving explicitly for problem size from $T_s = \frac{E}{1-E}T_0$ would give the necessary growth rate for maintaining constant E. This rate is called the isoefficiency function $I(P)$.

Problems like the 2-D DWT do not have simple closed-form expressions for $I(P)$. We therefore base the analysis on the following approach. Every term in T_0 of the four algorithms described in Section 4 is individually balanced with respect to (w.r.t.) the serial time T_s. Each term, in turn, establishes a lower bound on problem size growth rate. We define as problem size the parameter M^2 (total number of elements in input matrix $\mathbf{A}$), and express scalability as $M^2 = I(P)$.

In addition to the lower bound on isoefficiency implied by individual terms, there is a minimum growth rate imposed by the partitioning scheme which is independent of the algorithms, called *degree of concurrency* [23], arising due to the upper-bound on the maximum number of processors that can be used for solving a size-M^2 problem. In CP, the number of processors is upper-bounded by $P = O(M^2)$ (one element of $\mathbf{A}$ per processor), and in SP by $P = O(M)$ (one row of $\mathbf{A}$ per processor). These two expressions imply that the lower bound on isoefficiency imposed by concurrency is $M^2 = \Omega(P)$ for CP, and $M^2 = \Omega(P^2)$ for SP.

5.2. Isoefficiency Function Derivation

The parallel running times of NS-CP, S-CP and NS-SP, can all be written as two components distinguished by the condition $J > j_0$ or $J \le j_0$ (see for example Eqn. (12)). In the following analysis, it is only necessary to consider the component for which $J > j_0$. We omit the proof of this claim since it is simple but lengthy. Briefly, it rests on the fact that since J is assumed to scale with input size as $\lceil \gamma \log M \rceil$, it can then be shown that for some (M, P) pair, $J > j_0$ holds as long as $M = O(P^{\frac{1}{2(1-\gamma)}})$. This upper bound on input size is much larger than all the algorithm's isoefficiency functions derived in the following sections, and as a result, does not affect $I(P)$.

5.2.1. NS-CP

Overhead T_0 can be written as the product PT_p^{NS-CP} minus $T_s = t_c C^{NS}$. Since it is not possible to cancel the terms in PT_p^{NS-CP} that account for most of the computations (terms *(a)* and *(g)*) by simply subtracting T_s, we treat every term of PT_p^{NS-CP} shown in Table 5 as a candidate for an isoefficiency component. Those terms that should balance with T_s will imply no conditions on $I(P)$. This provides a simple rule in cases where it is difficult (or impossible) to algebraically cancel terms in PT_p by subtracting T_s: to treat every term in the product PT_p as if it where a component of T_0. Those terms that do not contribute to T_0 will not impose conditions on the isoefficiency when balanced w.r.t. T_s.

Consider term *(i)* of NS-CP on a CT-routed Mesh (Table 5) with associated constant $\beta_i = 4Lt_c + 2t_s + 2(L-1)t_w$. In order to maintain non-decreasing efficiency, and since

$T_0^{(i)} = \beta_i J P = \beta_i \gamma P \log M$, the following relation must hold

$$\frac{\beta_i \gamma P \log M}{T_s} = O(U), \tag{14}$$

where $U = (1 - E)/E$.

From Eqn. (7) we know that the serial time T_s is given by $\frac{8}{3} L t_c M^2 (1 - \frac{1}{4^J})$. We need only consider the dominant positive term M^2, as $T_s = \frac{8}{3} L t_c (M^2 - M^{2(1-\gamma)})$, and $0 < \gamma < 1$; as a result $T_s = \Theta(\beta_s M^2)$ with $\beta_s = 8 L t_c / 3$. The contribution of term (i) in NS-CP to the isoefficiency function is obtained by solving for M^2 in

$$\frac{\beta_i \gamma P \log M}{\Theta(\beta_s M^2)} = O(U), \tag{15}$$

which can be written as $\frac{M^2}{\log M} = \Omega(\frac{\beta_i \gamma P}{U \beta_s})$. By the degree of concurrency in CP we know that $M^2 = \Omega(P)$, implying that $M^2 = \Omega(\frac{\beta_i \gamma}{2U \beta_s} P \log P)$.

In Table 6 we list the growth rate of M^2 implied by each individual term in NS-CP. The constants β associated with each term are as given in Table 5. In interpreting Table 6 the reader should take into account the distinction between similarly labeled constants pertaining to different algorithms. For instance, constant β_g for algorithm NS-CP is *not* the same as β_g for S-CP; the former is equal to $2t_h - 4L t_c - 2(L - 1)t_w$, while the latter is equal to $2t_h$ (see Table 5, row (g)). Note that the constant for the NS-form serial execution time T_s^{NS} is $\beta_s = \frac{8}{3} L t_c$, and for the S-form $\beta_s = 4 L t_c$ (see terms (a), (b) and (c) in Table 5). Terms in T_p not inducing a lower bound on M^2 are not listed in Table 6.

5.2.2. S-SP, S-CP and NS-SP

According to pure asymptotic analysis, the S-SP algorithm on a Mesh (CT- and SF-routed) is unscalable. The reason is that term (m) in Table 5 leads to $T_0^{(m)} = \beta_m M^2 \sqrt{P}$, and there exists no relationship between M and P which will prevent this term from growing at a rate faster than $T_s = \Theta(\beta_s M^2)$. However, this asymptotically unscalable algorithm (S-SP) can still be as efficient as a scalable one (S-CP) over a range of network and problem size parameters. In section 5.3 and 6.1 we use the constants of the dominant terms as a basis for comparison to establish this range.

Despite the fact that S-SP on a CT-routed Mesh is asymptotically unscalable because there exist some P_0 beyond which it is not possible to maintain a given E, S-SP can still be characterized by a limited form of scalability over the range $1 \le P < P_0$. Consider the overall expression for S-SP's overhead, given by

$$T_0 = 2t_s P^{1.5} + t_w \sqrt{P} M^2 - 2t_s P - t_w M^2. \tag{16}$$

According to the relation $T_0/T_s = U$, and $T_s = 4 L t_c M^2$, one can solve for M^2 as

$$M^2 = \frac{2t_s (P^{1.5} - P)}{4 L t_c U + t_w (1 - \sqrt{P})}. \tag{17}$$

Table 6. Isoefficiency growth rates implied by individual terms in T_p, as labeled in Table 5, on a CT-routed Mesh. Functions shown represent growth rates of M^2. Note that similarly labeled constants differ among algorithms, and that $\beta_s = \frac{8}{3}Lt_c$ for the NS-form and $\beta_s = 4Lt_c$ for S-form.

	Term	NS-CP	S-SP	S-CP	NS-SP
(d)	$(1-\frac{1}{2^{j_0}})\frac{M}{\sqrt{P}}$	$(\frac{\beta_d}{U\beta_s})^2 P$			
(e)	$(1-\frac{1}{2^{j_0}})M$				$\frac{\beta_e}{U\beta_s}P^2$
(f)	$(\frac{1}{2^{j_0}}-\frac{1}{2^J})M$				$\frac{\beta_f}{U\beta_s}P^2$
(g)	j_0	$\frac{\beta_g}{2U\beta_s}P\log P$		$\frac{\beta_g}{2U\beta_s}P\log P$	$\frac{\beta_g}{2U\beta_s}P\log P$
(h)	$j_0\frac{M}{\sqrt{P}}$			$(\frac{\beta_h}{2U\beta_s})^2 P\log^2 P$	
(i)	J	$\frac{\beta_i\gamma}{2U\beta_s}P\log P$		$\frac{\beta_i\gamma}{2U\beta_s}P\log P$	
(j)	$J\frac{M}{\sqrt{P}}$			$(\frac{\beta_j\gamma}{2\beta_s U})^2 P\log^2 P$	
(k)	2^{J-j_0+1}	$(\frac{2\beta_k}{U\beta_s})^{\frac{2}{3-\gamma}} P^{\frac{3}{3-\gamma}}$		$(\frac{\beta_k}{U\beta_s})^{\frac{2}{3-\gamma}} P^{\frac{3}{3-\gamma}}$	$(\frac{2\beta_k}{U\beta_s})^{\frac{2}{3-\gamma}} P^{\frac{4}{3-\gamma}}$
(l)	$\sqrt{P}$		$\frac{\beta_l}{U\beta_s}P^{1.5}$		
(m)	$\frac{M^2}{\sqrt{P}}$		unscalable		
	concurrency	P	P^2	P	P^2

Let P_0 be the number of processors where S-SP becomes unscalable. Then, from Eqn. (17), M is unbounded when

$$P_0 = \left(\frac{4Lt_cU}{t_w} + 1\right)^2. \tag{18}$$

Since in S-SP the fastest isoefficiency rate is dictated by the degree of concurrency, its scalability may be approximated by the rate $M^2 = \Omega(P^2)$ over interval $1 \leq P < P_0$.

The individual contribution of every term to S-CP and NS-SP's isoefficiency $I(P)$, for a CT-routed Mesh, are also given in Table 6.

5.3. *Discussion*

After having identified the components in the isoefficiency function, we now proceed with the analysis of NS-CP's scalability, and with a comparison among the scalability of the two standard DWT forms: S-SP and S-CP. In Section 5.3.1 we establish that, although for large enough P NS-CP is scalable as $M^2 = \Omega(P^{\frac{3}{3-\gamma}})$, for realistic machines it is scalable as $M^2 = \Omega(P \log P)$. In Section 5.4 we make a similar statement regarding S-CP, and also give the range of P where S-SP is likely to be as efficient as S-CP.

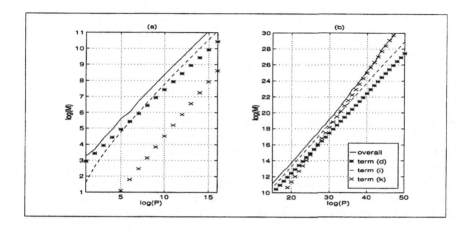

Figure 5. Overall isoefficiency of NS-CP on a CT-routed Mesh shown with solid line. Also shown are the contributions of individual terms *(d)*, *(i)* and *(k)*, in (a) $1 \le P \le 2^{16}$, and (b) $2^{15} \le P \le 2^{50}$.

5.3.1. NS-CP

NS-CP's term *(k)* in Table 6 induces the fastest asymptotic growth rate for problem size, i.e., $M^2 = \Omega(P^{\frac{3}{3-\gamma}})$. Term *(i)* implies a growth rate of $M^2 = \Omega(P \log P)$; and term *(d)* and the degree of concurrency a rate of $M^2 = \Omega(P)$. Term *(g)*, with a rate of $M^2 = \Omega(P \log P)$, has a negative constant, and can therefore be neglected. A term with a negative contribution to T_0 can only improve the isoefficiency function, and is necessarily implied by some other term with a positive constant.

Consider the following example. In Fig. 5 we show a plot of the terms in NS-CP's isoefficiency function, for $t_c = t_w = 1$, $t_s = 10$, $t_h = 0.1$, $E = 80\%$, $\gamma = 0.8$ and $L = 10$; Fig. 5 (a) shows the range $1 \le P \le 2^{16}$, and Fig. 5 (b) the range $2^{15} \le P \le 2^{50}$. Term *(d)* is shown with symbol '*', *(k)* with 'x', and *(i)* with a dashed line; P and M are restricted to powers of two, and shown in a $\log_2$ scale. Fig. 5 (a) shows that, although term *(k)* grows faster than terms *(i)* and *(d)*, for any practical number of processors, term *(k)* is very small in comparison with terms *(i)* and *(d)*. The solid line in this plot shows the *overall* isoefficiency function including the contribution from all three terms, and it is seen to closely follow the rate of the dominant term *(i)* for realistic P (Fig. 5 (a)). (The constant difference between overall isoefficiency and term *(i)* is due to the neglected term *(g)*.) The overall isoefficiency function was generated with a MATLAB-based simulation of the parallel computation times T_p. Given an efficiency level, E, in this simulation we iteratively compute the smallest problem size M achieving E for every P in $1 \le P \le 2^{16}$, and in $2^{15} \le P \le 2^{50}$.

Next, we determine the range where NS-CP is scalable as $M^2 = \Omega(P \log P)$ by considering the intersection of term *(k)* with terms *(i)* and *(d)*. Let P_i and P_d be the number of processors where term *(k)* is equal to terms *(i)* and *(d)*, respectively. Clearly, the region where term *(k)* is dominant is lower bounded by P_i and P_d. Equating term *(k)* and term *(d)*, we find that

$$P_d = \left(\frac{\beta_d}{U\beta_s}\right)^{\frac{2(3-\gamma)}{\gamma}} \left(\frac{U\beta_s}{2\beta_k}\right)^{\frac{2}{\gamma}}. \tag{19}$$

Assuming that $\gamma \to 1$, an approximate lower bound for P_d is $P_d > \frac{9L^2}{4U^2}(\frac{t_w^2}{t_c t_h})^2$. Similarly, an approximate lower-bound for P_i is given by $P_i > (2Lt_c + t_s + Lt_w)^{\frac{3-\gamma}{\gamma}}(\frac{1}{4t_h})^{\frac{2}{\gamma}}$.

In these two lower bounds t_h appears in the denominator, and, since hop-times are commonly one to two orders of magnitude smaller than the other three network parameters, one can expect large values for P_i and P_d. For instance, using the numbers from above example, we get $P_i > 2^{17.9}$ and $P_d > 2^{18.4}$. [9] Therefore, although term *(k)* does determine asymptotic scalability of NS-CP, one is not likely to encounter this growth rate in networks with less than 2^{18} processors, even in the worst case. And as a result, the scalability of NS-CP can be regarded as lower-bounded by the isoefficiency function $M^2 = \Omega(P \log P)$. Fig 5 (b) shows that overall isoefficiency is actually dominated by term *(k)* for $P > 2^{35}$.

Given that the functional form of NS-CP's scalability is $P \log P$, we now determine a general result regarding the effect of network and processor parameters on NS-CP's performance (efficiency). With appropriate substitutions, term *(i)* can be approximated as $\frac{3\gamma}{8U}\{2 + \frac{t_s + t_w}{t_c}\}P \log P$, and term *(d)* as $\frac{9}{U^2}(\frac{t_w}{t_c})^2 P$, where we see that communication parameters t_w and t_s both appear in the numerator, and processor period t_c in the denominator. This fact points to the well-known tradeoff between processor and network speed: That improving raw computational throughput might suggest increasing processor speed (reducing t_c). However, above two expressions indicate that if increased processor speed is not met by an increase in network speed, the result is a decrease in efficiency, which cancels the gain in processor speed.

In order to achieve the right combination of processor and network speed, there should be a balance among t_c, t_w and t_s. Even though these are not variable quantities on a particular machine, it is nonetheless very important for both, software and a hardware designers, to be aware of the basic limitation—and the ideal balance—between processor and network speed. Consider an infinitely fast network, N_∞, defined as one with $t_w, t_s, t_h \to 0$, and finite t_c.[10] It is easy to see that term *(i)* reduces to $\frac{3\gamma}{4U} P \log P$, and term *(d)* vanishes. The isoefficiency of N_∞ represents, then, the absolute lower bound for NS-CP on *any* real network. Consequently, one way of measuring the degree of balance among processor and network speed for an algorithm is in terms of its deviation from the efficiency of N_∞.

5.4.   S-SP and S-CP

To understand the potential usefulness of S-SP, we now summarize the scalability results for S-CP and compare the scalability of the two standard-form DWT algorithms: S-SP and S-CP. Similar reasoning as in Section 5.3.1 shows that S-CP terms *(g)* and *(k)* are not dominant for practical P, that term *(i)* is implied by term *(j)* for most network parameters, and that term *(h)* may be neglected since it has a negative constant. And as a result, S-CP is scalable as $M^2 = \Omega(P \log^2 P)$.

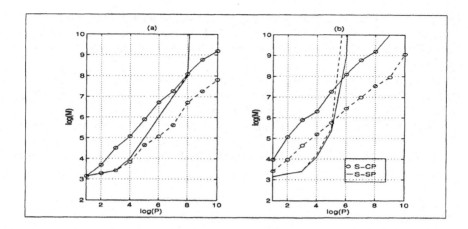

Figure 6. Overall isoefficiency of S-SP and S-CP on a CT-routed Mesh with $t_c = t_w = 1$, $t_s = 10$, $t_h = 0.1$ and $L = 10$. (a) $E = 0.7$, (b) $E = 0.85$. S-CP is shown with symbol 'o', and S-SP with no symbol. Solid lines show both algorithms with $\gamma = 0.95$, and dotted lines with $\gamma = 0.5$.

Fig. 6 shows the overall isoefficiency function of S-CP including the contribution from all terms (shown with symbol 'o'), and that of S-SP (no symbol), with parameters $t_c = t_w = 1$, $t_s = 10$, $t_h = 0.1$ and $L = 10$. In Fig. 6 (a), $E = 0.7$; and in Fig. 6 (b), $E = 0.85$. On both plots, the solid lines represent the isoefficiency curves evaluated with $\gamma = 0.95$; while the dotted lines $\gamma = 0.5$. In this example, S-SP shows improvement in performance over S-CP in its limited scalability range ($1 \le P < P_0$), and this improvement is larger for deep decompositions ($\gamma = 0.95$), where P_0 is given by Eqn. (18). Note that with $E = 0.7$, P_0 in Eqn. (18) evaluates to $329 = 2^{8.3}$, and with $E = 0.85$ evaluates to $65 \approx 2^6$, both quantities consistent with the plots (in (a), the last point on the S-SP curve is 2^8, and in (b), it is 2^6).

These plots underscore the difference in communication mechanisms among the two algorithms. In S-CP, communication overhead is strongly proportional to decomposition depth J, since a problem requiring many octaves will tend to rely on long distance communications among processors. The constant in the dominant term *(j)* of S-CP being proportional to γ^2 reflects this behavior. In contrast, in S-SP communication overhead is largely *independent* of decomposition depth J, because the algorithm incurs in exactly the same communication overhead for any J. Therefore, S-SP tends to exhibit larger performance gain relative to S-CP on problems with deep decomposition depth ($\gamma \to 1$). This is evident in Fig. 6, where S-SP curves with different γ are nearly identical; while dotted S-CP curves ($\gamma = 0.5$) are lower than the solid S-CP curves ($\gamma = 0.95$). Clearly, as γ increases, S-SP becomes more efficient than S-CP by merging communication requirements in a single phase. The main source of overhead in S-CP comes from octaves requiring long distance communications.

Efficiency gain of S-SP over S-CP with deep decompositions can be substantial. In Fig. 6 (b) we see that with $P = 16$ processors and $\gamma = 0.95$, S-CP requires an input matrix **A** with at least *sixteen* times as many elements in order to achieve the same efficiency as S-SP. However, as P_0 is inversely proportional to efficiency (via constant U in the expression for P_0, Eqn. (18)), the range where S-SP is scalable diminishes for high efficiency, becoming insignificant in this example for $E > 0.95$. As will be shown in Section 6, the performance gain of S-SP over S-CP is considerably higher on the Hypercube, where S-SP *is* a scalable algorithm.

It is also noteworthy that network start-up time t_s does not determine the range where S-SP is scalable. Hence, the relationship between the two isoefficiency curves for S-SP and S-CP, with very large t_s, is similar to that shown in Fig. 6. This is especially relevant in MIMD networks characterized by long start-up times. The critical ratio determining the potential gain of S-SP over S-CP is $U L t_c / t_w$ (see Eqn. (18)), dictating that fast processors on networks with relatively low channel bandwidth tend to lower P_0, diminishing the practical scalability range of S-SP.

5.5. NS-SP

Finally, the scalability of NS-SP can be shown to be dominated by terms that grow as $M^2 = \Omega(P^2)$ starting from very small values of P. Similar arguments as in Section 5.3.1 indicate that overall isoefficiency of NS-SP is given by $M^2 = \Omega(P^2)$, showing no improvement over NS-CP, S-CP or S-SP.

6. Hypercube and SF-routed Mesh

In this section, we briefly give the parallel execution times of the four algorithms and their isoefficiency functions on a Hypercube-connected computer and on a SF-routed Mesh. We also discuss similar scalability issues as in Section 5.3 on these networks.

6.1. CT-routed Hypercube

A $\sqrt{P} \times \sqrt{P}$ Mesh network with $P = 2^d$ can be embedded into a d-dimensional Hypercube with dilation and congestion of one, by applying a *Gray Coding* (GC) on the indices of the Mesh processors. The details of this well-known embedding may be found in [34]. An embedding with dilation and congestion of one preserves all neighbor-to-neighbor communication links present in the Mesh, and, as a result, parallel running times for the four algorithms on the Hypercube are lower bounded by the corresponding parallel times on the Mesh derived in Section 4.

In addition to preserving all Mesh links, GC embeddings on a Hypercube provide an extra set of links that might be used for improving efficiency. Although we have found no new communication schemes for making use of the extra links afforded by the hypercube, it can be shown that there is a negligible difference between the GC embedding lower bound of NS-CP and S-CP's isoefficiency function and that of the algorithms on a *fully-connected* network. A fully-connected network has a direct link among every single pair of processors. Therefore, the maximum possible improvement for the two checkerboard-partitioned algorithms, NS-CP and S-CP, in the hypothetical case that the extra Hypercube

Table 7. Relation between start-up time
t_s and t_c, t_h for given P' on a CT-routed
Hypercube with $\gamma = 0.95$, $L = 10$ and
$E = 0.9$.

P'	$t_s < \alpha(t_c + t_w + t_w^2/4t_c)$
	α
2048	4.79
512	12.84
128	31.09
32	63.45

links can be used, is nearly zero. Parallel execution times of NS-CP and S-CP on the Hypercube can be, then, taken to be equal to those on the Mesh.

The parallel running time of S-SP, however, improves considerably on a CT-routed Hypercube, and may be shown to be given by

$$T_p^{S-SP} = \frac{4t_c M^2}{P}\left(1 - \frac{1}{2^J}\right) + \left(t_s + t_w\left(\frac{M}{P}\right)^2\right)P + \frac{1}{2}t_h P \log P, \tag{20}$$

and its isoefficiency by two terms called $(p)M^2 = \Omega(\frac{t_s}{4ULt_c}P^2)$, and (q) $M^2 = \Omega(\frac{t_h}{8ULt_c}P^2 \log P)$. Unlike on the CT-routed Mesh, S-SP on the CT-routed Hypercube is scalable.

Next, we determine the range where the isoefficiency of S-SP is smaller than S-CP on the CT-routed Hypercube. Let P' be the number of processors where, for a given E, the problem sizes of S-SP and S-CP are equal; P' is the crossover point of the two isoefficiency functions given E. For determining a conservative lower bound on P', we consider term (j) of S-CP, and term (p) of S-SP. (From Section 5.4 we know that S-CP's term (j) is dominant, and one can easily show that S-SP's term (p) is larger than (q) for the range of interest, mostly due to t_h appearing in the numerator of (q).) Equating these two terms, it can shown that a lower bound on P' is implicitly given by the equation $\frac{P}{\log^2 P} < \tilde{\beta}$, where $\tilde{\beta} = \frac{\gamma^2 L(t_c + t_w/2)^2}{Ut_c t_s}$. Therefore, the condition on network parameters t_c, t_s, t_w that must be satisfied for a minimum desired P' is

$$t_s < \frac{\log^2 P'}{P'}\frac{\gamma^2 L}{U}(t_c + t_w + t_w^2/4t_c). \tag{21}$$

For illustration, consider the example in Fig. 6, where $t_c = t_w = 1$, $\gamma = 0.95$, $L = 10$ and $E = 0.85$ (note that t_h irrelevant). For a lower bound on P' of 128 processors we have that $t_s < 19.6(t_c + t_w + t_w^2/4t_c)$. In Table 7 we show the relation between t_s, t_c and t_w for $32 \leq P' \leq 2048$, with $L = 10$, $\gamma = 0.95$ and $E = 0.9$. Notice that typical values for P' on the Hypercube are considerably larger than P_0 on the Mesh as given in Section 5.4.

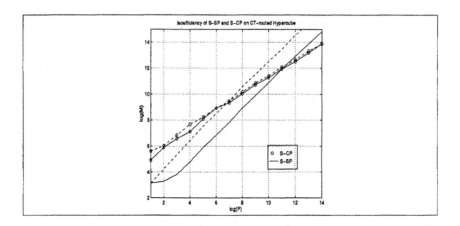

Figure 7. Isoefficiency of S-CP and S-SP on a CT-routed Hypercube, where $t_c = t_w = 1$, $t_h = 0.1$, $\gamma = 0.95$, $E = 0.9$ and $L = 10$. The solid lines show S-SP and S-CP with $t_s = 10$, while the dotted lines show S-SP and S-CP with $t_s = 100$.

Although there is no closed-form expression for P', it is clear that P' is proportional to $\frac{\gamma^2 L}{U t_s}(t_c + t_w + t_w^2/4t_c)$, implying that: (i) P' is directly proportional to γ, since, as $\gamma \to 1$, the condition on network parameters is relaxed, a fact which is consistent with the behavior observed on the CT-routed Mesh (Section 5.4); (ii) while P' is inversely proportional to t_s, P_0 for S-SP on the Mesh is independent of t_s; and (iii) P' on the CT-routed Hypercube is directly proportional to efficiency E.

In Fig. 7 we show the isoefficiency curves for S-CP (symbol 'o') and S-SP (no symbol), with $t_c = t_w = 1$, $t_h = 0.1$, $\gamma = 0.95$, $E = 0.9$ and $L = 10$. In this plot we can see the effect of varying start-up times. The solid curves show both S-SP and S-CP for $t_s = 10$; while the dotted curves for $t_s = 100$. The crossover point between S-SP and S-CP with $t_s = 10$ (solid lines) is seen to be at $P' = 2^{11}$, which is in agreement with Table 7, where in the first line we expect that, for $P' = 2048$ processors, a value for t_s of at least 10.77 is required. In Fig. 7 we can see that even with high start-up to computation time ratios (of over 100), S-SP is still likely to be as efficient as S-CP for less than 128 processors; note for instance that with $t_s = 100$ (dotted lines) P' occurs at 2^7.

6.2. SF-routed Mesh

The parallel execution times of NS-CP and S-CP on a SF-routed Mesh have almost the same functional form as on the CT-routed Mesh, except that the constants differ. Table 8 lists the constants in front of every term in the T_p expressions for NS-CP and S-CP, including β_r corresponding to a new term in S-CP, namely $\beta_r 2^{J-j_0} \frac{M}{\sqrt{P}}$. (S-SP is not shown, as its parallel running time on the SF-Mesh is the same as on the CT-Mesh.)

Scalability terms for NS-CP, S-SP, and S-CP on the SF-routed Mesh are derived from

Table 8. Constants for NS-CP and S-CP's execution times T_p on a SF-routed Mesh. Note new term (r) in S-CP. (T_p of S-SP is the same as on the CT-routed Mesh.)

	Term	NS-CP	S-CP
(a)	$(1 - \frac{1}{4j_0})\frac{M^2}{P}$	$\frac{8}{3}Lt_c$	
(c)	$(1 - \frac{1}{2j_0})\frac{M^2}{P}$		$4Lt_c$
(d)	$(1 - \frac{1}{2j_0})\frac{M}{\sqrt{P}}$	$4(L-2)t_w$	
(g)	j_0	$4t_s + 2t_h - 4Lt_c$	$2t_h$
(h)	$j_0\frac{M}{\sqrt{P}}$		$2(L-2)t_w - 4Lt_c$
(i)	J	$4Lt_c + 2t_s$	$2t_s$
(j)	$J\frac{M}{\sqrt{P}}$		$4Lt_c$
(k)	2^{J-j_0+1}	$2(L-1)t_w + 2t_h$	$2t_h$
(r)	$2^{J-j_0}\frac{M}{\sqrt{P}}$		$4(L-1)t_w$

the T_p expressions of Table 8, and are omitted for reasons of space. A similar analysis as in Section 5.3.1 shows that the intersection of scalability term (k) with terms (i) and (g) of NS-CP on the SF-routed Mesh can be approximated by the lower bound on P of $P > (2t_s - 2Lt_c)^{\frac{3-\gamma}{\gamma}}(\frac{1}{4(L-1)t_w})^{\frac{2}{\gamma}}$. In contrast to the scalability of NS-CP on the CT-routed Mesh, the scalability of NS-CP on the SF-routed Mesh is dominated by term $M^2 = \Omega\left((\frac{2\beta_k}{U\beta_s})^{\frac{2}{(3-\gamma)}} P^{\frac{3}{(3-\gamma)}}\right)$, but over a much *smaller* range than on the CT-routed Mesh; in fact, the lower bound on P can be very close to $P = 1$ for typical numbers, implying that NS-CP's scalability is dominated by this term over all P. As $\gamma \to 1$, this asymptotic scalability expression becomes $M^2 = \Omega(P^{3/2})$, a considerably faster rate than on the CT-routed Mesh where isoefficiency is $M^2 = \Omega(P \log P)$.

The scalability of S-CP can be shown to be dominated by the new term (r), implying a growth on input size of $M^2 = \Omega\left((\frac{\beta_r}{U\beta_s}P)^{\frac{2}{(2-\gamma)}}\right)$. As $\gamma \to 1$, this rate becomes $M^2 = \Omega(P^2)$. One can show that the isoefficiency function of S-SP is lower than that of S-CP over the range $1 \leq P < P_0$, with *higher* efficiency gain margin than in the CT-routed Mesh case of Section 5.4, and with P_0 as given in Eqn. (18). The reason for this higher efficiency gain margin is that, while S-SP's performance is identical on the SF- and CT-routed Mesh, S-CP's is poorer on the SF-routed Mesh as compared to the CT-routed Mesh. As a result, the same conclusion as in the CT-routed Mesh case of Section 5.4 also holds on the SF-routed Mesh, namely that S-SP is as efficient as S-CP over the range $1 \leq P < P_0$.

Table 9. Summary of overall isoefficiency growth rates of M^2. P is the number of processors, M^2 the number of elements in the input matrix, and γ is a parameter relating M to the total number of octaves J, as $J = \lceil \gamma \log M \rceil$.

Network	NS-CP	S-SP	S-CP	NS-SP
CT Hypercube	$\Omega(P \log P)$	$\Omega(P^2)$	$\Omega(P \log^2 P)$	$\Omega(P^2)$
CT Mesh	$\Omega(P \log P)$	unscalable	$\Omega(P \log^2 P)$	$\Omega(P^2)$
SF Mesh, Hypercube	$\Omega\left(P^{\frac{3}{3-\gamma}}\right)$	unscalable	$\Omega\left(P^{\frac{2}{2-\gamma}}\right)$	$\Omega(P^2)$

7. Conclusions

In this paper we have established the scalability properties of four 2-D DWT parallel algorithms on the Mesh and Hypercube networks with both CT- and SF-routing. We have considered the Standard and Non-standard DWT algorithm forms in conjunction with checkerboard and row stripped partitioning. Understanding the scalability properties of an algorithm on a particular network has important practical implications because scalability quantifies a parallel system's ability for making efficient use of increasing computational resources—i.e., adding processors to an existing machine. In Table 9 we summarize the *practical* scalability of our four algorithms by listing their isoefficiency growth rate of M^2; that is, for almost any practical machine size available in today's technology (tens of thousands of processors), these algorithms are characterized by the isoefficiency terms in Table 9 despite the fact that they may have faster growing terms. Note that S-SP is only asymptotically scalable on the CT-routed Hypercube; in all other networks there exists no problem size that will maintain constant efficiency for any P.

In addition to deriving the asymptotic scalability of the four 2-D DWT algorithms, we have also presented a more detailed analysis based on the proportionality constants involved in the isoefficiency expressions. This analysis allows us to define the *practical scalability* of an algorithm, that is, its scalability over a realistic range of machine and problem sizes. In particular, asymptotics alone would suggest that, on the CT-routed Hypercube, the Standard form with stripped partitioning (S-SP) is considerably less scalable than its alternative version with checkerboard partitioning (S-CP), the former scalable as P^2, and the latter as $P \log^2 P$. However, detailed analysis shows that for any number of processors less than a threshold P', S-SP is actually *more* efficient than S-CP; the threshold P' being proportional to $\frac{\gamma^2 L}{t_s}(t_c + t_w + t_w^2/4t_c)$, where t_c, t_w and t_s are the normalized computation time of one multiply-add operation, the network word-transfer and the start-up times respectively, and L is the wavelet filter length. For typical parallel machines, P' can be in the thousands of processors. The S-SP algorithm has advantages over its checkerboard partitioned alternative in terms of its implementation and portability, making it an attractive solution. A similar result showing that S-SP is more efficient that S-CP on the CT- and SF-routed Mesh also holds, but over a smaller range of processors, and with threshold P_0 being proportional to the ratio Lt_c/t_w, and inversely proportional to efficiency. In all processor networks, it was seen that S-SP takes advantage of its ability to merge all communications into a single phase, particularly in problems requiring deep decompositions, or large γ. We also found that the Non-standard form with stripped partitioning (NS-SP) shows no performance advantage over the other three algorithms for any network and problem size parameters.

Acknowledgements

This work was partially supported by the Advanced Projects Reserach Agency of the Department of Defense under grants MDA-972-31-0023 and R49620-93-1-0490 monitored by the Air Force Office of Scientific Research.

Notes

1. An algorithm-network pair is usually called a parallel system.

2. $S_G^{j_1, j_2}$ is introduced simply to distinguish S from S_G as the coarsest component $j_1, j_2 = J$.

3. In the rest of the paper, all log functions are base two.

4. Toroidal connections are not necessary, unless a DWT with circular convolution and edge wrap-around is sought. We do not assume toroidal connections, but their inclusion is straight-forward.

5. Due to decimation by two, only $L - 2$ elements are transferred and not $L - 1$.

6. This assumption may be relaxed, but general P simply obscures main discussion points.

7. With correct memory allocation and deallocation, structure $\tilde{I}$ may occupy the same memory space of I' because, at the end of $\mathbf{F}_{comm}$, PE_x may relinquish the space occupied by I'.

8. A system is said to be linearly scalable if the input must grow as a *linear* function of the number of processors to maintain equal performance.

9. P_l and P_d are typically much larger quantities, as common ratios of t_c, t_s, t_w to hop-time t_h are much larger than 10. However, for this argument it is sufficient to find simple closed-form lower bound expressions.

10. N_∞ closely resembles a PRAM machine.

References

1. A. Grossmann and R. Kronland-Martinet, and J. Morlet, "Reading and Understanding Continuous Wavelet Transforms". in J. M. Combes, A. Grossmann, and P. Tchamitchian (eds.), *Wavelets: Time-Frequency Methods and Phase Space*, pp. 2–20. Springer-Verlag, 1989.

2. I. Daubechies, *Ten Lectures on Wavelets*, Number 61 in CBMS–NSF Series in Applied Mathematics. Philadelphia: SIAM Publications, 1992.

3. P. P. Vaidyanathan, *Multirate Systems and Filter Banks*, Englewood Cliffs, NJ: Prentice-Hall, 1993.

4. S. G. Mallat, "A Theory for Multiresolution Signal Decomposition: The Wavelet Representation," *IEEE Transactions on Pattern Analysis and Machine Intelligence*, vol. 11, no. 7, 1989, pp. 674–693.

5. M. Vetterli and C. Herley. "Wavelets and Filter Banks: Theory and Design," *IEEE Trans. on Signal Processing*, vol. 40, no. 9, 1992, pp. 2207–2232.

6. R. A. DeVore, B. Jawerth, and B. J. Lucier, "Image Compression Through Wavelet Transform Coding," *IEEE Transactions on Information Theory*, vol. 38, no. 2, 1992, pp. 719–746.

7. W. Zettler, J. Huffman, and D. C. P. Linden, "Applications of compactly supported wavelets to image compression," Technical Report AD 900119, Aware Inc., 1990.

8. M. Unser and A. Aldroubi, "Polynomial Splines and Wavelets—A Signal Processing Perspective," in C. K. Chui (ed.), *Wavelets: A Tutorial in Theory and Applications*. Academic Press, 1992.

9. G. Beylkin, R. Coifman, and V. Rokhlin, "Fast Wavelet Transforms and Numerical Algorithms I," *CPAM*, vol. 44, 1991, pp. 141–183.

10. B. Alpert, "A Class of Bases in L^2 for the Sparse Representation of Integral Operators," *SIAM Journal of Scientific Computation*, vol. 24, no. 1, 1993, pp. 246–262.

11. B. Alpert, G. Beylkin, R. Coifman, and V. Rokhlin, "Wavelets for the Fast Solution of Second-Kind Integral Equations," *SIAM Journal on Scientific Computation*, vol. 14, no. 1, 1993, p. 159.

12. E. Bacry, S. Mallat, and G. Papanicolaou, "A Wavelet Based Space-Time Adaptive Numerical Method for Partial Differential Equations," *Mathematical Modeling and Numerical Analysis*, vol. 26, 7, 1992, p. 793.

13. S. Mallat and W. L. Hwang, "Singularity detection and processing with wavelets," Courant Institute of Mathematical Sciences, New York University, Preprint.

14. E. H. Adelson, E. Simoncelli, and R. Hingorani, "Orthogonal Pyramid Transforms for Image Coding," in *Visual Communications and Image Processing II*, no. 845, pp. 50–58. SPIE, 1987.

15. T. Bell, "Remote Sensing," *IEEE Spectrum*, 1995, pp. 25–31.

16. M. Hamdi, "Parallel Architectures for Wavelet Transforms," in *1993 Workshop on Computer Architecures for Machine Perception*, 1993, pp. 376–384.

17. C. Chakrabarti and M. Vishwanath, "Efficient Realizations of the Discrete and Continuous Wavelet Transforms: from Single Chip Implementations to Mappings on SIMD Array Computers," *IEEE Transactions on Signal Processing*, vol. 43, no. 3, 1995, pp. 759–771.

18. J. Fridman and E. Manolakos, "On the Synthesis of Regular VLSI Architectures for the 1-D Discrete Wavelet Transform," in A. Lane and M. Unser (eds.), *Wavelet Applications in Signal and Image Processing II*, pp. 91–104. SPIE, July 1994.

19. J. Fridman and E. S. Manolakos, "Distributed Memory and Control VLSI Architectures for the 1-D Discrete Wavelet Transform," in J. Rabaey, P. Chau, and J. Eldon (eds.), *VLSI Signal Processing VII*, pp. 388–397. IEEE Signal Processing Society, 1994.

20. J. Fridman and E. S. Manolakos, "Discrete Wavelet Transform: Data Dependence Analysis and Synthesis of Distributed Memory and Control Architectures," *IEEE Transactions on Signal Processing*, to appear 1997.

21. J. Fridman, *Parallel Algorithms and Architectures for the Discrete Wavelet Transform*, PhD thesis, Northeastern University, June 1996.

22. J. Fridman, B. York, and E. S. Manolakos, "Discrete Wavelet Transform Algorithms on a Massively Parallel Platform," in *International Conference on Signal Processing Applications and Technology*, pp. 1512–1516, 1995.

23. V. Kumar, A. Grama, A. Gupta, and G. Karypis, *Introduction to Parallel Computing*, The Benjamin/Cummings Publishing Company, Inc., 1994.

24. J. P. Andrew, P. O. Ogunbona, and F. J. Paoloni, "Comparison of Wavelet Filters and Subband Analysis Structures for Still Image Compression," in *Proc. Int'l. Conf. on Acoustics, Speech and Signal Proccessing*, pp. V–589–592. IEEE Signal Processing Society, 1994.

25. G. A. Geist, M. T. Heath, B. W. Peyton, and P. H. Worley, "PICL: A portable Instrumented Communication Library," Technical Report, ORNL/TM-11130, Oak Ridge National Laboratory, 1990.

26. H. Xu, E. T. Kalns, P. K. McKinley, and L. M. Ni, "ComPaSS: A Communication Package for Scalable Software Design," *Journal of Parallel and Distributed Computing*, vol. 22, 1994, pp. 449–461.

27. S. Y. Kung, *VLSI Array Processors*, Prentice Hall, 1989.

28. J. L. Gustafson, "Reevaluating Amdahl's Law," *Communications of the Association for Computing Machinery (ACM)*, vol. 31, no. 5, 1988, pp. 532–533.

29. P. H. Worley, "The Effect of Time Constraints on Scaled Speedup," *SIAM Journal Sci. Statist. Comput.*, vol. 11, no. 5, 1990, pp. 838–858.

30. X. H. Sun and L. M. Ni, "Another View of Parallel Speedup," in *Proceedings Supercomputing 90*, pp. 324–333, 1990.

31. F. A. Van-Catledge, "Toward a General Model for Evaluating the Relative Performance of Computer Systems". *Int'l. Journal of Supercomputer Applications*, vol. 3, 1989, pp. 100–108.

32. X. H. Sun and D. Rover, "Scalability of Parallel Algorithm-Machine Combinations," *IEEE Transactions on Parallel and Distributed Systems*, vol. 5, no. 6, 1994, pp. 599–613.

33. V. Kumar and A. Gupta, "Analyzing Scalability of Parallel Algorithms and Architectures," *Journal of Parallel and Distributed Computing*, vol. 22, 1994, pp. 379–391.

34. J. Ghosh, S. K. Das, and A. John, "Concurrent Processing of Linearly Ordered Data Structures in Hypercube Multicomputers," *IEEE Transactions on Parallel and Distributed Systems*, vol. 5, no. 9, 1994, pp. 898–911.

Multidimensional Systems and Signal Processing, 8, 219–227 (1997)

A Fast Algorithm to Map Functions Forward

WAYNE LAWTON

Institute of Systems Science, National University of Singapore, Kent Ridge, Singapore 119597

Abstract. Mapping functions forward is required in image warping and other signal processing applications. The problem is described as follows: specify an integer $d \geq 1$, a compact domain $D \subset R^d$, lattices L_1, $L_2 \subset R^d$, and a deformation function $F : D \to R^d$ that is continuously differentiable and maps D one-to-one onto $F(D)$. Corresponding to a function $J : F(D) \to R$, define the function $I = J \circ F$. The forward mapping problem consists of estimating values of J on $L_2 \cap F(D)$, from the values of I and F on $L_1 \cap D$. Forward mapping is difficult, because it involves approximation from scattered data (values of $I \circ F^{-1}$ on the set $F(L_1 \cap D)$), whereas backward mapping (computing I from J) is much easier because it involves approximation from regular data (values of J on $L_2 \cap D$). We develop a fast algorithm that approximates J by an orthonormal expansion, using scaling functions related to Daubechies wavelet bases. Two techniques for approximating the expansion coefficients are described and numerical results for a one dimensional problem are used to illustrate the second technique. In contrast to conventional scattered data interpolation algorithms, the complexity of our algorithm is linear in the number of samples.

Key Words: image warping, interpolation kernels and subdivision, scattered data interpolation, Daubechies orthonormal wavelet basis, approximate expansions using moments

1. Introduction

Let $d \geq 1$ be an integer and let $D \subset R^d$ be a compact domain. A *deformation* (on D) is a continuously differentiable function $F : D \to R^d$ that maps D one-to-one onto $F(D)$. Let F be a deformation, let $F^{-1} : F(D) \to D$ denote the inverse of F, and let $I : D \to R$ and $J : F(D) \to R$ be a pair of functions that satisfies

$$I = J \circ F. \tag{1.1}$$

Equation (1.1) is summarized by the statement that the following diagram of functions *commutes* (where id denotes the identity mapping of R).

$$
\begin{array}{ccc}
D & \xrightarrow{\ F\ } & F(D) \\
{\scriptstyle I}\downarrow & & \downarrow{\scriptstyle J} \\
R & \xrightarrow[\ \text{id}\ ]{} & R
\end{array}
$$

We say that F maps (or transforms) the function I *forward* to J and that F maps the function J *backward* to I. Equation (1.1) is equivalent to the equation

$$J = I \circ F^{-1}, \tag{1.2}$$

therefore, backward mapping is simpler than forward mapping because it is obtained from composing functions whereas forward mapping requires the additional operation of inverting the function F.

Mapping functions in signal processing arises when F represents a warping of the time domain ($d = 1$) of an acoustic signal induced by motion between the source and receiver. We briefly discuss the role of mapping functions in image science where the spatial domain ($d = 2$) of a *continuous image* I is mapped forward to a function J by a function F that can represent pin cushion or barrel distortion, encountered in vidicon cameras and cathode ray tube displays, or perspective distortion, resulting when an extended object is viewed from an oblique angle ([18], p. 429).

The process of mapping image functions forward (or backward) (also called *image warping* and *geometric image transformation*), so as to obtain J from I (or I from J), is clearly required to correct geometric image distortions. It has many other applications that are extremely useful in medical imaging, remote sensing, geometric modelling, computer graphics and animation, pattern recognition and machine vision, see ([22], pp. 1–5) and ([17], pp. 1–4).

The first use of mapping image functions is described in a landmark paper by a group of eminent Michigan scientists [3] who developed optical systems to map images. More recent related developments are described in [9].

Although optical systems have enormous spatio-temporal bandwidths, and therefore are capable of processing high resolution images with great speed, they lack both flexibility and control. The evolution of digital image sensors and digital computing has been accompanied by the rise of digital image processing that uses image samples consisting of the values of the continuous image at a set of lattice points in the domain. The images are assumed to be sufficiently well behaved and the lattice is assumed to be sufficiently *fine* (have closely spaced points) so that the continuous image can be accurately interpolated from its samples. A more complete discussion of continuous images and image sampling is given in ([18], pp. 1–24, 93–120).

Digital image mapping requires two operations ([8], p. 297):

- computing the deformation F (or its inverse F^{-1}),

- computing sample values of J (or I)

This paper addresses the second operation, called *image resampling* in the literature ([22], pp. 117–161). Section 2 reviews interpolation theory. Section 3 describes a general method for digital forward mapping that approximates the function J by a truncated orthonormal expansion $\widetilde{J}$ that uses compactly supported basis functions constructed from Daubechies scaling functions and wavelet functions. Two techniques are described for approximating the integrals that represent the expansion coefficients. The first technique uses a change of variables to replace the unknown function J in each integrand by the known function I. The second technique replaces the unknown function J in each integrand by a polynomial that approximates J over the support of the corresponding basis function. Section 4 describes numerical results using the second technique for a simple experiment where $d = 1$ and the basis functions consist of translates of Daubechies 8 coefficient scaling function.

2. Interpolation

Let $d \geq 1$ be an integer. A *lattice* is an additive subgroup $L \subset R^d$ spanned by d linearly independent vectors, hence $L = MZ^d$ for some invertible $d \times d$ matrix M. The *meshsize* Δ of a lattice L is defined to be $2/\sqrt{d}$ times the maximum distance between $x \in R^d$ and L (hence the meshsize of the lattice Z^d equals 1).

Let $m, n \geq 0$. A compactly supported function $K : R^d \to R$ is called an *interpolation kernel* with *smoothness order* m and *approximation order* n if

* $K(0) = 1$ and $K(p) = 0$ for all $p \in Z^d \backslash \{0\}$,

* $K \in C^m$ (space of m times continuously differentiable functions),

* $\sum_{p \in Z^d} Q(p) K(x - p) = Q(x)$ whenever Q is a polynomial of degree $< n$.

Let K be an interpolation kernel and let $L = MZ^d$ be a lattice with meshsize Δ. Let $f : R^d \to R$ be a continuous function and construct the function $\tilde{f} : R^d \to R$ by

$$\tilde{f}(x) \equiv \sum_{p \in L} f(p) K(M^{-1}[x - p]). \tag{2.3}$$

The summation converges since, for each $x \in R^d$, only a finite number of terms are nonzero. The first condition above ensures $\tilde{f}(p) = f(p)$, for all $p \in L$, thus $\tilde{f}$ interpolates f on L. The second condition ensures $\tilde{f} \in C^m$. The third condition implies that if $f \in C^n$ then $\tilde{f}$ approximates f with error bounded by a constant times Δ^n. This condition is equivalent to the Strang-Fix conditions [21] of order n:

$$\widehat{K}(0) = 0 \text{ and } D^j \widehat{f}(2\pi q) = 0 \text{ whenever } q \in Z^d \backslash \{0\} \text{ and } 0 \leq j \leq n - 1. \tag{2.4}$$

Riemenschneider and Shen recently constructed families of compactly supported piecewise polynomial lattice interpolation kernels having arbitrarily high smoothness and approximation order [20]. The support of these kernels increases approximately linearly with m and n, thus providing an opportunity to obtain higher interpolation accuracy with modest increases in computational complexity. Stationary subdivision schemes [1] also provide efficient lattice interpolation methods.

Let L_1, L_2 be lattices and assume values of a deformation F on $L_1 \cap D$ are given. We consider two *digital mapping problems*.

* Backward Mapping Problem: approximate values of I on a subset $A \subset L_1 \cap D$ from values of J on $L_2 \cap F(D)$,

* Forward Mapping Problem: approximate values of J on a subset $B \subset L_2 \cap F(D)$ from values of I on $L_1 \cap D$.

We assume A and B exclude a sufficient number of lattice points close to the boundary of

D. Equation (1.1) and the assumption that J can be accurately interpolation from its values on $L_2 \cap F(D)$ imply

$$I(p) = J(F(p)) \approx \widetilde{J}(F(p)) \equiv \sum_{q \in L_2 \cap F(D)} J(q)K(M^{-1}[F(p) - q]); \text{ for } p \in A, \qquad (2.5)$$

where $K(x)$ is a suitable interpolation kernel. Therefore, digital backward mapping reduces to a problem of interpolation from regular data (values of J on $L_2 \cap F(D)$).

Digital forward mapping is equivalent to computing values of J on B, from values of J on the set $F(L_1 \cap D)$. Direct computation requires *scattered data interpolation* whose complexity increases faster than linearly with the number of samples, see [16], because a different kernel function must be computed for each sample value.

REMARK 1 Fast interpolation algorithms exist for special cases of scattered data. For example, synthetic aperture radar and computer aided tomography systems may provide data samples on a concentric squares polar grid. A fast algorithm for interpolating from these grids can be obtained from the algorithm developed in ([11]). However, this algorithm requires $O(N \log N)$ operations per output sample (where N is the number of points in A) compared to $O(N)$ operations required by lattice interpolation.

An alternate forward mapping approach computes values of F^{-1} on B, and subsequently uses equation (1.2) to obtain

$$J(p) = I(F^{-1}(p)) \approx \sum_{q \in L_1 \cap D} I(q)K(M^{-1}[F^{-1}(p) - q]) \text{ for } p \in B. \qquad (2.6)$$

This approach reduces forward mapping to a backward mapping problem. However, computing F^{-1} may be difficult. If F is expressed by a nonlinear polynomial, by Shepard interpolation, or by a thin plate spline, then F^{-1} does not admit a closed form and each value of F^{-1} requires solving two simultaneous nonlinear equations. Iterative methods, such as Newton's method, must be used. The speed of these methods is limited by the requirement of finding a sufficiently close initial approximation. If F is not given explicitly but only values of F on $L_1 \cap D$ are given, then F^{-1} may be computed using surface fitting or using scattered data interpolation methods as described in [19] and [10].

3. Orthonormal Expansion Method

The general orthonormal expansion method approximates J on the set $B \subset L_2 \cap F(D)$ by

$$J(u) \approx \widetilde{J}(u) \equiv \sum_i \widetilde{g}_i(q)G_i(u) \qquad (3.7)$$

where G_i is a family of compactly supported orthonormal functions and each $\widetilde{g}_i$ approximates the expansion coefficient

$$g_i \equiv \int J(y)G_i(y)dy. \qquad (3.8)$$

We briefly describe two techniques to compute $\widetilde{g}_i$. The first technique substitutes $y = F(x)$ and equation (1.1) $J(F(x)) = I(x)$ to obtain

$$g_i = \int I(x) G_i(F(x))) \left| \det \frac{\partial F}{\partial x} \right| dx. \tag{3.9}$$

For each G_i, the function F is approximated by an affine function over the support of $G_i(F(x))$ and I is approximated by a polynomial over the support of $G_i(F(x))$. Then moments of G_i are used to compute $\widetilde{g}_i$.

The second technique approximates $J(y)$ over the support of each $G_i(y)$ by a polynomial $Q_i(y)$ determined from values $J(F(p)) = I(p)$ for a subset of $p \in L_1 \cap D$ such that $F(p)$ is near the support of G_i. Then moments of G_i are used to compute $\widetilde{g}_i$.

REMARK 2 The use of basis functions G_i having the form $G(y - q)$, $q \in L_2 \cap F(D)$ where G is constructed from Daubechies' scaling functions [5], [6] is advantageous for several reasons. First, each $G(y - q)$ can be expressed, using the refinement equation, as a linear combination of translates of a function having arbitrarily small support to ensure the affine and polynomial approximations are accurate and thus the approximate expansion coefficients $\widetilde{g}(q)$ are accurate. Furthermore, G can be selected to have arbitrarily high regularity and approximation order to ensure good truncated approximations. Finally, closed expressions for moments of G and for values of G at rational linear combinations of elements in L_2 can be used to facilitate computations as in [7].

REMARK 3 If G is the simplest function constructed from a scaling function, the *box car function* (characteristic function of a square), then the application of the refinement equation with the first technique yields the *splitting-shooting* and related *splitting-integrating* image mapping methods developed in ([17], pp. 95–150).

4. Numerical Results

A simple numerical experiment was performed to illustate the second technique to approximate the expansion coeficients. Here $d = 1$, $L_1 = L_2 = Z$, $D = [0, 255]$, and

$$F(x) = x + .2x(255 - x)/256, \quad I(x) = sin(2\pi x/64).$$

The basis functions were $G(y - q)$, $q \in L_2 \cap D$ where G is the Daubechies 8 coefficient scaling function. Thus G has support interval $[0, 7]$ and moments of order 0 thru 3

$$moments = (1.0000, 1.0054, 1.0108, .9074),$$

and values at integers 1 thru 6

$$values = (1.0072, -.0338, .0396, -.0118, -.0012, .0000)$$

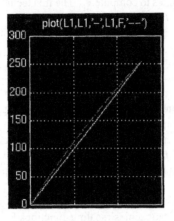

Figure 1. Mapping function F.

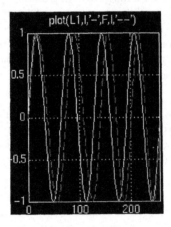

Figure 2. Image I and mapped function J.

The closed form for the inverse mappping function

$$F^{-1}(x) = 256(1.2 - \sqrt{1.44 - .8x/256})/.4$$

was used to compute the exact values $J(q) = I(F^{-1}(q))$ for $q \in L_2 \cap D$. Each approx-imating polynomial Q_q was a cubic determined to minimize the squared error over a set of six values of $I(F(p))$ for $F(p)$ near q. The moments of G were used to compute the approximate expansion coefficients and the integer values of G were then used to compute an approximation Ja of J on $L_2 \cap D$.

Figure 1 uses a dashed plot to display the mapping function F on D and solid plot to display the identity function on D. Figure 2 displays the *image* function I (represents an

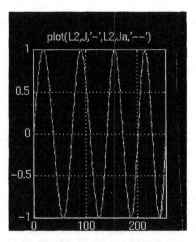

Figure 3. J and Approximation Ja.

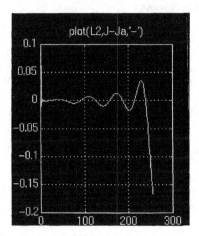

Figure 4. Error function $J - Ja$.

image for $d = 2$) by a solid plot and displays the mapped function J by a dashed plot. Figure 3 compares overlays the plots of the function J and its approximation Ja while Figure 4 displays the error $J - Ja$. Clearly the error increases with the *instantaneous frequency* of the mapped function J and increases sharply near the right boundary of D due to the truncation of the expansion.

REMARK 4 Further theoretical and experimental work is required to quantify the potential advantages the orthonormal expansion method may have over the simpler use of local polynomial approximation as in the second technique to compute the values directly on each point in $B \subset L_2 \cap F(D)$. We conjecture that the advantages would include anti aliasing properties (similar to those enjoyed by the Galerkin method of discretization as opposed to the collocation method) and more robustness if F is nearly non invertible. Other areas for research include the use of symmetric complex valued wavelets to mitigate boundary effects as described in [15].

References

1. A. S. Cavaretta, W. Dahmen, and C. A. Micchelli, "Stationary Subdivision," *Memoirs of the American Mathematical Society*, vol. 93, 1991, pp. 1–186.

2. A. Cohen, I. Daubechies, and J. C. Feauveau, "Biorthogonal bases of compactly supported wavelets," *Communications on Pure and Applied Mathematics*, vol. 45, 1992, pp. 485–560.

3. L. J. Cutrona, E. N. Leith, C. J. Palermo, and L. J. Porcello, "Optical data processing and filtering systems," *IRE Transactions on Information Theory*, vol. IT-6, 1960, pp. 384–400.

4. W. Dahmen and C. A. Micchelli, "On stationary subdivision and the construction of compactly supported orthonormal wavelets," *Numerical Mathematics*, vol. 94, 1990, pp. 69–89.

5. I. Daubechies, "Orthonormal bases of compactly supported wavelets," *Communications on Pure and Applied Mathematics*, vol. 41, 1988, pp. 909–996.

6. I. Daubechies, *Ten Lectures on Wavelets*, Philadelphia: Society for Industrial and Applied Mathematics, 1992.

7. R. Glowinski, W. Lawton, M. Ravachol, and E. Tenenbaum, "Wavelet solution of linear and nonlinear elliptic, parabolic, and hyperbolic problems in one space dimension," in R. Glowinski and A. Lichnewsky (eds.), *Proceedings of the Ninth International Conference on Computing Methods in Applied Sciences and Engineering*, Paris, France, 1990. SIAM, Philadelphia.

8. R. C. Gonzales and R. E. Woods, *Digital Image Processing*, Reading, MA: Addison-Wesley, 1992.

9. J. L. Horner, *Optical Signal Processing*, New York: Academic Press, 1987.

10. C. L. Lawson, "Software for C^1 Surface Interpolation," *Mathematical Software III*, J. R. Rice (ed.), London: Academic Press, 1977, pp. 161–194.

11. W. Lawton, "A new polar Fourier transform for computer-aided tomography and spotlight synthetic aperture radar," *IEEE Transactions on Acoustics, Speech, and Signal Processing*, vol. 36, no. 6, 1988, pp. 931–933.

12. W. Lawton, "Necessary and sufficient conditions for constructing orthonormal wavelet bases," *Journal of Mathematical Physics*, vol. 32, no. 1, 1991, pp. 57–61.

13. W. Lawton, "Multilevel properties of the wavelet-Galerkin operator," *Journal of Mathematical Physics*, vol. 32, no. 6, 1991, pp. 1440–1443.

14. W. Lawton, "Application of complex-valued wavelets to subband decomposition," *IEEE Transactions on Signal Processing*, vol. 41, no. 12, 1993, pp. 3566–3568.

15. W. Lawton, S. L. Lee, and Z. Shen, "Characterization of compactly supported refinable splines," *Advances in Computational Mathematics*, vol. 3, 1995, pp. 137–145.

16. R. Franke and G. Nielson, "Scattered Data Interpolation and Applications: A Tutorial and Survey," H. Hagen and D. Roller (eds.), *Geometric Modelling: Method and Applications*, New York: Springer-Verlag, 1991, pp. 131–159.

17. Z. C. Li, T. D. Bui, Y. Y. Tang and C. Y. Suen, *Computer Transformations of Digital Images and Patterns*, Singapore: World Scientific, 1989.

18. W. K. Pratt, *Digital Image Processing*, New York: John Wiley, 1978.

19. D. Ruprecht and H. Müller, "Image warping with scattered data interpolation methods," in H. Hagen and D. Roller (eds.), *Geometric modelling: Methods and applications*, New York: Springer-Verlag, 1991.

20. S. D. Riemenschneider and Z. Shen, "General interpolation on the lattice hZ^s: Compactly supported fundamental solutions," *Numerische Mathematik*, to appear.

21. G. Strang and G. Fix, "A Fourier analysis of the finite element variational method," in G. Geymonat (ed.), *Constructive Aspects of Functional Analysis*, C.I.M.E., 1973, pp. 739–840.

22. G. Wohlberg and T. E. Boult, "Digital Image Warping," *IEEE Computer Society Press*, 1990.

Multidimensional Systems and Signal Processing, 8, 229–238 (1997)
© 1997 Kluwer Academic Publishers, Boston.

Contributing Authors

Sankar Basu received B.Sc and B.Tech degrees from the Calcutta University respectively in Physics and Radio Physics and electronics, M.S and Ph.D degrees in electrical engineering from the University of Pittsburgh. He has been a faculty member at Stevens Institute of Technology, NJ. He has visited the Lehrstuhl fur the NAchrichtentechnik at the Ruhr Inversitat, bochum, Germany as an Alexander von Humboldt fellow for an extended period of time, and was a visiting research scientist at the Information & Decision Systems Laboratory at MIT. In addition, he has been a senior summer research scientist at the Naval Underwater Systems Center in Connecticut. Currently, he is with the IBM research at the T. J. Watson research center. His research interests has been signal processing with particular emphasis on multidimensional signals, and the mathematics of networks and systems theory. Most recently he has been involved in speech recognition research.

Steven A. Benno received his B.S. degree from Rutgers University in 1987, his M.S. from Columbia University in 1989, and is currently a Ph.D. candidate at Carnegie Mellon University for electrical engineering.

From 1987 to 1991, he was a Member of Technical Staff at AT&T Bell Laboratories doing work in adaptive processing, and automatic detection and classification of ocean acoustic signals. From 1989 to 1991, he was also an adjunct professor at the New Jersey Institute of Technology.

His current research interests include time-frequency, time-scale representations, feature extraction, and pattern recognition. He is a member of Tau Beta Pi and Eta Kappa Nu.

Giancarlo Calvagno was born in Venezia, Italy, in 1962. He received the "Laurea in Ingegneria Eletronica" degree from the University of Padova in 1986, and the Doctorate degree in electrical engineering from the University of Padova in 1990. From 1988 to 1990 he was at the Coordinated Science Laboratory of the University of Illinois at Urbana-Champaign as a Visiting Scholar. Since 1990 he has been with the Dipartimento di Elettronica e Informatica of the University of Padova, as a Ricercatore. His main research interests are in the areas of digital signal processing, signal reconstruction, image processing and coding.

José Fridman received the Ph.D. degree in Electrical and Computer Engineering from Northeastern University in 1996, and the B.S. and M.S. degrees both in Electrical Engineering from Boston University in '87 and '91, respectively. His primary research focus is on the computational aspects of signal processing and numerical algorithms, and his research interests include parallel processing, communication networks and fault tolerant computing. Dr. Fridman is currently with Analog Devices, Inc., and may be reached by E-mail at Jose.Fridman@analog.com.

Ton A. C. M. Kalker was born in The Netherlands in 1956. He received his M.S. degree in mathematics in 1979 from the University of Leiden, The Netherlands. From 1979 until 1983, while he was a Ph.D. candidate, he worked as a Research Assistant at the University of Leiden. From 1983 until December 1985 he worked as a lecturer at the Computer Science Department of the Technical University of Delft. In January 1986 he received his Ph.D. degree in Mathematics.

In December 1985 he joined the Philips Research Laboratories Eindhoven. Until January 1990 he worked in the field of Computer Aided Design. He specialized in (semi-) automatic tools for system verification. Currently he is a member of the Digital Signal Processing group of Philips Research, where he is working on advanced methods for video compression. His research interests include wavelets, multirate signal processing, motion estimation, psycho physics and digital video compression.

Bernard C. Levy received the diploma of Ingénieur Civil des Mines from the Ecole National Supérieure des Mines in Paris, France, and the Ph.D. in Electrical Engineering from Stanford University. From July 1979 to June 1987 he was Assistant and then Associate Professor in the Department of Electrical Engineering and Computer Science at M.I.T. Since July 1987, he has been with the University of California at Davis, where he is currently Professor of Electrical Engineering and a member of the Graduate Group in Applied Mathematics. He was a consultant with the Charles Stark Draper Laboratory in Cambridge, Massachusetts from 1986 to 1989, and from January to July 1993, he was a Visiting Researcher at the Institut de Recherche en Informatique et Systèmes Aléatoires (IRISA), in Rennes, France.

His research interests are in multidimensional and statistical signal processing, estimation, detection, inverse problems, and scientific computing. He is a Fellow of IEEE, a member of SIAM and the Acoustical Society of America, and currently serves as a member of the Image and Multidimensional Signal Processing technical committee of the IEEE Signal Processing Society.

Michael Lightstone received the B.S. degree in electrical engineering from the University of Illinois at Urbana-Champaign in 1990 and the M.S. and Ph.D. degrees in electrical engineering from the University of California at Santa Barbara in 1992 and 1995, respectively.

During his undergraduate studies, he interned at a number of companies including McDonnell Douglas Aircraft Company in St. Louis, MO; AT&T Bell Laboratories in Naperville, IL; and Andersen Consulting in Chicago, IL. From 1991 to 1993 he was a visiting researcher at the Tampere University of Technology in Tampere, Finland; the Advanced Video Technology Department at AT&T Bell Laboratories in Murray Hill, NJ; and the Image Processing and Analysis Group at Jet Propulsion Laboratory in Pasadena, CA. He is currently with Chromatic Research, Inc., in Mountain View, CA. His research interests are in the areas of image and video compression and processing. He is a member of Eta Kappa Nu and Tau Beta Pi.

Eric E. Majani was born in Akron, OH, in 1960. He received the *Ingénieur* Degree from E.N.S.A.E. (Ecole Nationale Supérieure de l'Aéronautique et de l'Espace), and the *Diplôme d'Etudes Avancées (DEA)* Degree from the University Paul Sabatier, both in Toulouse, France, in 1982. He received the *M.Sc.* and *Ph.D.* degrees in Electrical Engineering from the California Institute of Technology, Pasadena, CA, respectively in 1983 and 1988. Since 1988, he has been a member of technical staff at the Jet Propulsion Laboratory (JPL), Pasadena, CA, mostly with JPL's Image Analysis Systems group. His current research interests include: image segmentation, image registration, reversible integer transforms, wavelet transform design, image compression and multiresolution image analysis.

Elias S. Manolakos (elias@cdsp.neu.edu) received the PhD degree in Computer Engineering from University of Southern Calfornia in 1989, the MSEE degree from University of Michigan, Ann Arbor, and the Diploma in Electrical Engineering from the National Technical University of Athens, Greece. He was the recipient of the Pan-Hellenic Award in Computer Science and Engineering of the Greek State Scholarships Foundation (IKY) in 1984, and of an NSF Research Initiation Award in 1993. Since 1989 Dr. Manolakos is with the Electrical and Computer Engineering Dept. of Northeastern University, where he is currently an Associate Professor and Associate Director of the Communications and Digital Signal Processing (CDSP)  Center for Research and Graduate Studies. He is leading the Parallel Processing & Architectures research group of CDSP and his research interests include: Parallel Computing for Signal/Image Processing, Systematic Synthesis of Parallel Array Algorithms and Architectures, Neural Networks for Signal Processing, and Fault Tolerant Computing. His research is being supported by the National Science Foundation and the Defense Advanced Research Project Agency.

Dr. Manolakos is a member of the IEEE Signal Processing Society's Technical Committees on the Design and Implementation of Signal Processing Systems and Neural Networks. He was the Program Chair of the 1995 IEEE International Workshop on Neural Networks for Signal Processing and has participated in the organizing and program committees for several other IEEE Conferences. He is an Associate Editor for the IEEE Transactions on Signal Processing and of the Journal of VLSI Signal Processing, Kluwer Acad. Publ. He has authored or co-authored more than 40 refereed publications and co-edited two books.

More information on his research activities is available on the World Wide Web URL http://www.cdsp.neu.edu/info/manolakos.html.

Thomas G. Marshall, Jr. received the B.S.E.E. and M.S.E.E. degrees from Purdue University in 1952 and 1955, respectively, and the Licenciate and Doctor of Technology degrees from Chalmers University of Technology, Gothenburg Sweden, in 1968 and 1976, respectively.

He served in the Signal Corps from 1955 to 1954 at White Sands Proving Ground and was commanding officer at Oscura Range Camp during this period. He was with the David Sarnoff Research Center from 1955 to 1966 working with transistor circuits for television applications for which he received seven patents and a joint Achievement Award for developing a transistorized, portable TV set. He joined Rutgers University in 1968 where he is currently a Professor of Electrical and Computer Engineering. He was department chair from 1976 to 1979 and founded the Center for Digital Signal Processing Research in 1987. His recent research interests have been in real-number error-correcting codes and digital filter banks and transmultiplexers.

Charles A. Micchelli graduated in 1969 from Stanford University with a Ph.D. in mathematics, and since 1970 has been a member of the research staff of the IBM Thomas J. Watson Research Center at Yorktown Heights, NY. He has held visiting and adjunct positions at numerous universities in the U.S., Europe, South America, and Israel and has lectured widely on his research. In 1983, he was invited to the International Congress of Mathematicians in Warsaw and in 1990 was a CBMS lecturer at Kent State University. He serves on the editorial board of nine mathematics journals, is the author of the book *Mathematical Aspects of Geometric Modeling*, published by SIAM in 1994, and has written or coauthored more than 180 papers. His research interests are in computational mathematics. In 1992, Dr. Micchelli received a Humboldt Award; in 1994, he was on sabbatical at RWTH-Aachen as a visiting professor.

photo
not
available
at time of
print

Eric L. Miller (S'90, M'95) received the S.B. in 1990, the S.M. in 1992, and the Ph.D. degree in 1994 all in Electrical Engineering and Computer Science at the Massachusetts Institute of Technology, Cambridge, MA.

He is currently an assistant professor in the Department of Electrical and Computer Engineering at Northeastern University. His research interests include the exploration of theoretical and practical issues surrounding the use of multiscale and statistical methods for the solution of inverse problems in general and inverse scattering problems in particular; development of computationally efficient, physically-based models for use in signal processing applications; multiscale methods for reduction of computational complexity.

Dr. Miller is a member of Tau Beta Pi, Eta Kappa Nu, and Phi Beta Kappa.

Sanjit K. Mitra received the B.Sc. (Hons.) degree in Physics in 1953 from Utkal University, Cuttack, India; the M.Sc. (Tech.) degree in Radio Physics and Electronics in 1956 from Calcutta University; the M.S. and Ph.D. degrees in Electrical Engineering from the University of California, Berkeley, in 1960 and 1962, respectively. From June 1962 to June 1965, he was at the Cornell University, Ithaca, New York, as an Assistant Professor of Electrical Engineering. He was with the AT&T Bell Laboratories, Holmdel, New Jersey, from June 1965 to January 1967. He has been on the faculty at the University of California since then, first at the Davis campus and more recently at the Santa Barbara campus as a Professor of Electrical and Computer Engineering, where he served as Chairman of the Department from July 1979 to June 1982. He served as the President of the IEEE Circuits and Systems Society in 1986. He is currently a member of the editorial boards of the *International Journal on Circuits, Systems and Signal Processing*, *Multidimensional Systems and Signal Processing*; *Signal Processing*; and the *Journal of the Franklin Institute*.

Dr. Mitra is the recipient of the 1973 F. E. Terman Award and the 1985 AT&T Foundation Award of the American Society of Engineering Education, the Education Award of the IEEE Circuits & Systems Society in 1989 and the Distinguished Senior U.S. Scientist Award from the Alexander von Humboldt Foundation of West Germany in 1989. In May 1987 he was awarded an Honorary Doctorate of Technology degree from the Tampere University of Technology, Tampere, Finland. Dr. Mitra is a Fellow of the IEEE, AAAS and SPIE, and a member of EURASIP and ASEE.

José M. F. Moura received the engenheiro electrotécnico degree in 1969 from Instituto Superior Tecnico (IST), Lisbon, Portugal, and the M.Sc., E.E., and the D.Sc. in Electrical Engineering and Computer Science from the Massachusetts Institute of Technology (M.I.T.), Cambridge, in 1973 and 1975, respectively.

He is presently a Professor of Electrical and Computer Engineering at Carnegie Mellon University (CMU), Pittsburgh, which he joined in 1986. Prior to this, he was on the faculty of IST where he was an Assistant Professor (1975), Professor Agregado (1978), and Professor Catedratico (1979). He has had visiting appointments at several Institutions, including M.I.T. (Genrad Associate Professor of Electrical Engineering and Computer Science, 1984–1986, also associated with LIDS) and the University of Southern California (research scholar, Department of Aerospace Engineering, Summers 1978–1981). His research interests lie in statistical signal processing (one and two dimensional), image and video processing, array processing, underwater accoustics, and multiresolution techniques. He has organized and codirected two international scientific meetings on signal processing theory and applications.

Dr. Moura has over 160 technical contributions, including invited ones, published in international journals and conference proceedings, and·is co-editor of the books *Nonlinear Stochastic Problems* (Reidal, 1983) and *Acoustic Signal Processing for Ocean Exploration* (Kluwer, 1993).

Dr. Moura is currently the Editor in Chief for the *IEEE Transactions in Signal Processing* and a member of the *IEEE Press Board*. He was elected Fellow of the IEEE in November 1993 and is a corresponding member of the *Academy of Sciences of Portugal* (Section of Sciences). He is affiliated with several IEEE societies, Sigma Xi, AMS, IMS, and SIAM.

Hyungju Park (also known as H. Alan Park) was born in Korea, and got the B.S. degree in physics from Seoul National University in 1986. Then he came to the graduate school of University of California at Berkeley to study mathematics. At Berkeley, he was a Teaching Assistant in the Department of Mathematics and a Research Assistant in the Department of Electrical Engineering and Computer Sciences.

For the first few years at Berkeley, most of the education and training he got was on algebraic geometry. But eventually, his research interest shifted to computational aspects of this classical field and their applications to various problems in multidimensional signal processing.

He got the Ph.D. degree in mathematics in 1995 under the supervision of his two research co-advisors: Tsit-Yuen Lam (mathematics) and Martin Vetterli (electrical engineering). He was a member of the Berkeley Wavelet Group led by Martin Vetterli and became an Assistant Professor of Mathematical Sciences at Oakland University, Rochester, Michigan, in the fall of 1995.

He is a member of IEEE and of AMS.

Roberto Rinaldo obtained the "Laurea in Ingegneria Elettronica" degree in 1987 from the University of Padova, Padova, Italy, with honors and the medal for the highest graduation score. He obtained the MS degree from the University of California at Berkeley, and the Doctorate degree from the University of Padova in 1992. Since 1993 he has been with the Dipartimento di Elettronica e Informatica of the University of Padova, where he is currently a Ricercatore in Communications and Signal Processing. His interests are in the field of multidimensional signal processing, video signal coding, fractal theory and image coding.

Martin Vetterli received the Dipl. El.-Ing. degree from ETH Zürich, Switzerland, in 1981, the MS degree from Stanford University in 1982, and the Doctorat ès Science degree from EPF Lausanne, Switzerland, in 1986.

He was a Research Assistant at Stanford and EPFL, and has worked for Siemens and AT&T Bell Laboratories. In 1986, he joined Columbia University in New York where he was last an Associate Professor of Electrical Engineering and co-director of the Image and Advanced Television Laboratory. In 1993, he joined the University of California at Berkeley where he is a Professor in the Dept. of Electrical Engineering and Computer Sciences. Since 1995, he is a Professor of Communication Systems at EPF Lausanne, Switzerland.

He is a fellow of the IEEE, a member of SIAM, and of the editorial boards of *Signal Processing, Image Communication, Annals of Telecommunications, Applied and Computational Harmonic Analysis* and *The Journal of Fourier Analysis and Applications*.

He received the Best Paper Award of EURASIP in 1984 for his paper on multidimensional subband coding, the Research Prize of the Brown Bovery Corporation (Switzerland) in 1986 for his thesis, and the IEEE Signal Processing Society's 1991 Senior Award for a 1989 Transactions paper with D. LeGall. He was a plenary speaker at the 1992 IEEE

ICASSP in San Francisco, and is the co-author, with J. Kovacevic, of the book *Wavelets and Subband Coding* (Prentice-Hall, 1995).

His research interests include wavelets, multirate signal processing, computational complexity, signal processing for telecommunications, digital video processing and compression and wireless video communications.

Alan S. Willsky received both the S.B. degree and the Ph.D. degree from the Massachusetts Institute of Technology in 1969 and 1973 respectively. He joined the M.I.T. faculty in 1973 and his present position is Professor of Electrical Engineering. From 1974 to 1981 Dr. Willsky served as Assistant Director of the M.I.T. Laboratory for Information and Decision Systems. He is also a founder and member of the board of directors of Alphatech, Inc. In 1975 he received the Donald P. Eckman Award from the American Automatic Control Council. Dr. Willsky has held visiting positions at Imperial College, London, L'Université de Paris-Sud, and the Institut de Recherche en Informatique et Systèmes Aléatoires in Rennes, France.

He was program chairman for the 17th IEEE Conference on Decision and Control, has been an associate editor of several journals including the *IEEE Transactions on Automatic Control*, has served as a member of the Board of Governors and Vice President for Technical Affairs of the IEEE Control Systems Society, was program chairman for the 1981 Bilateral Seminar on Control Systems held in the People's Republic of China, and was special guest editor of the 1992 special issue of the *IEEE Transactions on Information Theory* on wavelet transforms and multiresolution signal analysis. Also in 1988 he was made a Distinguished Member of the IEEE Control Systems Society. In addition Dr. Willsky has given several plenary lectures at major scientific meetings including the 20th IEEE Conference on Decision and Control, the 1991 IEEE International Conference on Systems Engineering, the SIAM Conf. on Control 1992, 1992 Inaugural Workshop for the National Centre for Robust and Adaptive Systems, Canberra, Australia, and the IEEE Symposium on Image and Multidimensional Signal Processing in Cannes, France in 1993.

Dr. Willsky is the author of the research monograph *Digital Signal Processing and Control and Estimation Theory* and is co-author of the undergraduate text *Signals and Systems*. He was awarded the 1979 Alfred Noble Prize by the ASCE and the 1980 Browder J. Thompson Memorial Prize Award by the IEEE for a paper excerpted from his monograph. Dr. Willsky's research interests are in the development and application of advanced methods of estimation and statistical signal and image processing. Methods he has developed have been successfully applied in a wide variety of applications including failure detection in high-performance aircraft, advanced surveillance and tracking systems, electrocardiogram analysis, computerized tomography, and remote sensing.

Yuesheng Xu is an assistant professor of mathematics. He graduated from Department of Computer Science Zhongshan University, Guangzhou, China, in 1982 with B.S. and in 1985 with M.S. and obtained his Ph.D. degree of Computational and Applied Mathematics from Old Dominion University, U.S.A., in 1989. His research interest ranges from approximation theory to numerical analysis. In particular, his current research interest is the wavelet construction and applications to numerical solutions of boundary integral equations. He has published approximately 40 research papers in three research areas: constrained best approximation, numerical solutions of integral equations with weakly singular kernels, and wavelet constructions and their applications. His research project "wavelet constructions on finite domains and applications in boundary integral equations" (co-principal investigator Charles A. Micchelli) is supported by the National Science Foundation of the United States. He has been a NASA research grant awardee since 1992 and was a NASA summer research fellow in 1995.